P9-BYR-614

PACKING FOR MARS

ALSO BY MARY ROACH

Stiff: The Curious Lives of Human Cadavers

Spook: Science Tackles the Afterlife

Bonk: The Curious Coupling of Science and Sex

PACKING FOR MARS

THE CURIOUS SCIENCE OF LIFE IN THE VOID

MARY ROACH

W. W. NORTON & COMPANY ✳ NEW YORK ✳ LONDON

Printed in the United States of America
First Edition

Photograph credits: p. 13: © Hamilton Sundstrand Corporation 2010. All rights reserved; p. 21: Image by Deirdre O'Dwyer; p. 41: Dmitri Kessel / Time & Life Pictures / Getty Images; p. 63: Courtesy of NASA; p. 79: CBS Photo Archive / Hulton Archive / Getty Images; p. 95: Courtesy of NASA; p. 107: Image Source / Getty Images; p. 129: Courtesy of NASA; p. 149: Bettman/Corbis; p. 173: Ryan McVay / Riser / Getty Images; p. 191: Courtesy of NASA; p. 209: Hulton Archive / Getty Images; p. 229: Joanna McCarthy / Riser / Getty Images; p. 247: Hulton Archive / Getty Images; p. 265: Courtesy of NASA; p. 285: Courtesy of NASA; p. 307: Tim Flach / Stone+ / Getty Images

For information about permission to reproduce selections from this book, write to Permissions, W. W. Norton & Company, Inc., 500 Fifth Avenue, New York, NY 10110

For information about special discounts for bulk purchases, please contact W. W. Norton Special Sales at specialsales@wwnorton.com or 800-233-4830

Manufacturing by RR Donnelley, Harrisonburg
Book design by Ellen Cipriano
Production manager: Julia Druskin

Library of Congress Cataloging-in-Publication Data

Roach, Mary.
Packing for Mars : the curious science of life in the void / Mary Roach.—1st ed.
p. cm.
Includes bibliographical references.
ISBN 978-0-393-06847-4 (hardcover)
1. Space biology—Popular works. I. Title.
QH327.R63 2010
571.0919—dc22

2010017113

W. W. Norton & Company, Inc.
500 Fifth Avenue, New York, N.Y. 10110
www.wwnorton.com

W. W. Norton & Company Ltd.
Castle House, 75/76 Wells Street, London W1T 3QT

1 2 3 4 5 6 7 8 9 0

For Jay Mandel and Jill Bialosky,
with cosmic gratitude

*

CONTENTS

PACKING FOR MARS

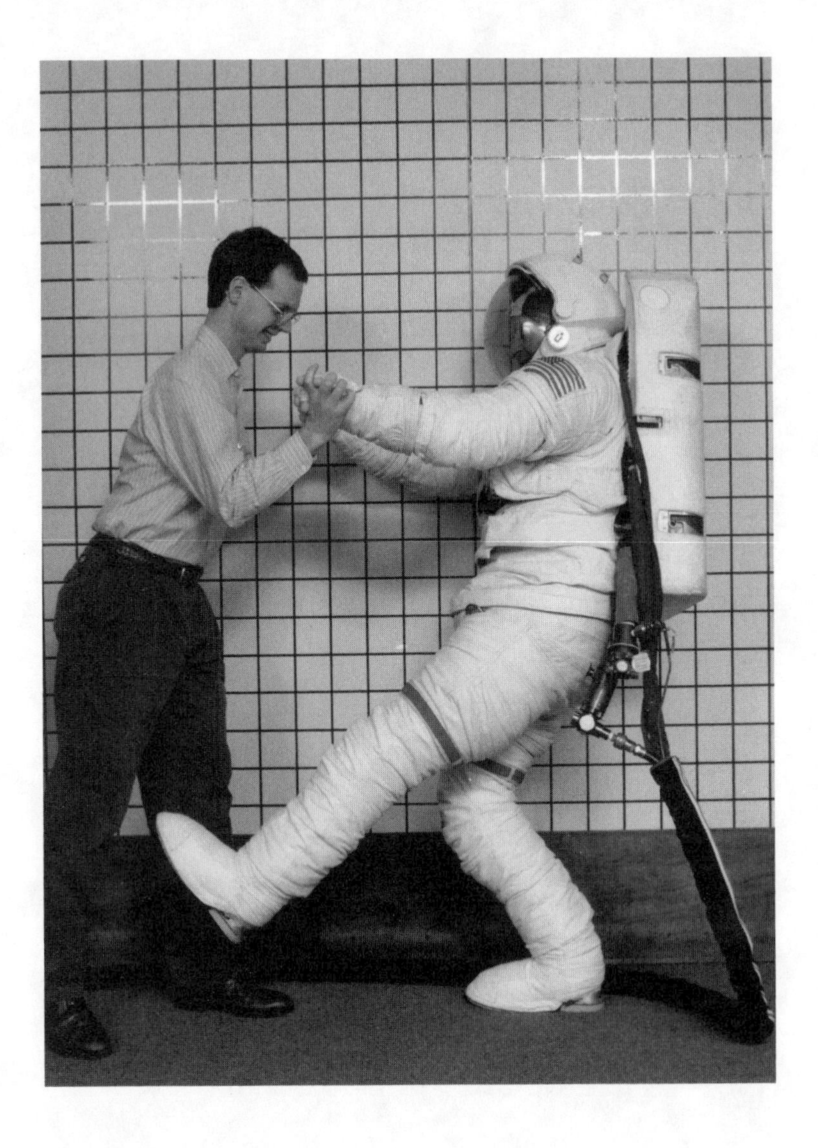

COUNTDOWN

To the rocket scientist, you are a problem. You are the most irritating piece of machinery he or she will ever have to deal with. You and your fluctuating metabolism, your puny memory, your frame that comes in a million different configurations. You are unpredictable. You're inconstant. You take weeks to fix. The engineer must worry about the water and oxygen and food you'll need in space, about how much extra fuel it will take to launch your shrimp cocktail and irradiated beef tacos. A solar cell or a thruster nozzle is stable and undemanding. It does not excrete or panic or fall in love with the mission commander. It has no ego. Its structural elements don't start to break down without gravity, and it works just fine without sleep.

To me, you are the best thing to happen to rocket science. The human being is the machine that makes the whole endeavor so endlessly intriguing. To take an organism whose every feature has evolved to keep it alive and thriving in a world with oxygen, gravity, and water, to suspend that organism in the wasteland of space for a month or a year, is a preposterous but captivating undertaking. Everything one takes for granted on Earth must be rethought, relearned, rehearsed—full-grown men and women toilet-trained, a chimpanzee dressed in a flight suit and launched into orbit. An entire odd universe of mock outer space has grown up here on Earth. Capsules that never blast off; hospital wards where healthy people spend months on their backs, masquerading zero gravity; crash labs where cadavers drop to Earth in simulated splash-downs.

A couple years back, a friend at NASA had been working on something over in Building 9 at the Johnson Space Center. This is the building with the mock-ups, some fifty in all—modules, airlocks, hatches, capsules. For days, Rene had been hearing an intermittent, squeaking racket. Finally, he went to investigate. "Some poor guy in a spacesuit running on a treadmill suspended from a big complicated gizmo to simulate Martian gravity. Lots of clipboards and timers and radio headsets and concerned looks all around." It occurred to me, reading his email, that it's possible, in a way, to visit space without leaving Earth. Or anyway, a sort of slapstick-surreal make-believe edition. Which is more or less where I've been these past two years.

OF THE MILLIONS of pages of documents and reports generated by the first moon landing, none is more telling, to me anyway, than an eleven-page paper presented at the twenty-sixth annual meeting of the North American Vexillological Association. Vexillology is the study of flags, not the study of vexing things, but in this case, either would fit. The paper is entitled "Where No Flag Has Gone Before: Political and Technical Aspects of Placing a Flag on the Moon."

It began with meetings, five months before the Apollo 11 launch. The newly formed Committee on Symbolic Activities for the First Lunar Landing gathered to debate the appropriateness of planting a flag on the moon. The Outer Space Treaty, of which the United States is a signer, prohibits claims of sovereignty upon celestial bodies. Was it possible to plant a flag without appearing to be, as one committee member put it, "taking possession of the moon"? A telegenically inferior plan to use a boxed set of miniature flags of all nations was considered and rejected. The flag would fly.

But not without help from the NASA Technical Services Division. A flag doesn't fly without wind. The moon has no atmosphere to speak of, and thus no wind. And though it has only about a sixth the gravity of Earth, that is enough to bring a flag down in an inglorious droop. So a crossbar was hinged to the pole and a hem sewn along the top of the flag. Now the Stars and Stripes would appear to be flying in a brisk wind—convincingly enough to prompt decades of moon hoax jabber—though in fact it was hanging, less a flag than a diminutive patriotic curtain.

Challenges remained. How do you fit a flagpole into the cramped, overpacked confines of a Lunar Module? Engineers were sent off to design a collapsible pole and crossbar. Even then, there wasn't room. The Lunar Flag Assembly—as flag, pole, and crossbar had inevitably come to be known—would have to be mounted on the outside of the lander. But this meant it would have to withstand the 2,000-degree Fahrenheit heat generated by the nearby descent engine. Tests were undertaken. The flag melted at 300 degrees. The Structures and Mechanics Division was called in, and a protective case was fashioned from layers of aluminum, steel, and Thermoflex insulation.

Just as it was beginning to look as though the flag was finally ready, someone pointed out that the astronauts, owing to the pressurized suits they'd be wearing, would have limited grip strength and range of motion. Would they be able to extract the flag assembly from its insulated sheath? Or would they stand there in the gaze of millions, grasping futilely? Did they have the reach needed to extend the telescoping segments? Only one way to know: Prototypes were made and the crew convened for a series of flag-assembly deployment simulations.

Finally came the day. The flag was packed (a four-step procedure supervised by the chief of quality assurance) and mounted on the Lunar Module (eleven steps), and off it went to the moon. Where the telescoping crossbar wouldn't fully extend and the lunar soil was so hard that Neil Armstrong couldn't plant the staff more than about 6 or 8 inches down, creating conjecture that the flag was most likely blown over by the engine blast of the Ascent Module.

Welcome to space. Not the parts you see on TV, the triumphs and the tragedies, but the stuff in between—the small comedies and everyday victories. What drew me to the topic of space exploration was not the heroics and adventure stories, but the very human and sometimes absurd struggles behind them. The Apollo astronaut who worried that he, personally, was about to lose the moon race for the United States by throwing up on the morning of his spacewalk, causing talk of tabling it. Or the first man in space, Yuri Gagarin, recalling that as he walked the red carpet before the Presidium of the Central Committee of the Communist Party of the Soviet Union and a cheering crowd of thousands, he noticed that his shoelace was undone and could think of nothing else.

At the end of the Apollo program, astronauts were interviewed to get their feedback on a range of topics. One of the questions: If an astronaut were to die outside the spacecraft during a spacewalk, what should you do? "Cut him loose," read one of the answers. All agreed: An attempt to recover the body could endanger other crew members' lives. Only a person who has experienced firsthand the not insignificant struggle of entering a space capsule in a pressurized suit could so unequivocally utter those words. Only someone who has drifted free in the unlimited stretch of the universe could understand that burial in space, like the sailor's burial at sea, holds not disrespect but honor. In orbit, everything gets turned on its head. Shooting stars streak past below you, and the sun rises in the middle of the night. Space exploration is in some ways an exploration of what it means to be human. How much normalcy can people forgo? For how long, and what does it do to them?

Early in my research, I came across a moment—forty minutes into the eighty-eighth hour of Gemini VII—which, for me, sums up the astronaut experience and why it fascinates me. Astronaut Jim Lovell is telling Mission Control about an image he has captured on film—"a beautiful shot of a full Moon against the black sky and the strato formations of the clouds of the earth below," reads the mission transcript. After a momentary silence, Lovell's crewmate Frank Borman presses the TALK button. "Borman's dumping urine. Urine [in] approximately one minute."

Two lines further along, we see Lovell saying, "What a sight to behold!" We don't know what he's referring to, but there's a good chance it's not the moon. According to more than one astronaut memoir, one of the most beautiful sights in space is that of a sun-illumined flurry of flash-frozen waste-water droplets. Space doesn't just encompass the sublime and the ridiculous. It erases the line between.

1

HE'S SMART BUT HIS BIRDS ARE SLOPPY

Japan Picks an Astronaut

First you remove your shoes, as you would upon entering a Japanese home. You are given a pair of special isolation chamber slippers, light blue vinyl imprinted with the Japan Aerospace Exploration Agency logo, the letters *JAXA* leaning forward as though rushing into space at terrific speed. The isolation chamber, a freestanding structure inside building C-5 at JAXA's headquarters in Tsukuba Science City, is in fact a home of sorts, for one week, for the ten finalists competing for two openings in the Japanese astronaut corps. When I came here last month, there wasn't much to see—a bedroom with curtained "sleeping boxes," and an adjoining common room with a long dining table and chairs. It's more about being seen. Five closed-circuit cameras mounted near the ceiling allow a panel of psychiatrists, psychologists, and JAXA managers to observe the applicants. To a large extent, their behavior and the panel's impressions of them during their stay will determine which two will wear the JAXA logo on spacesuits instead of slippers.

The idea is to get a better sense of who these men and women are, and how well they're suited to life in space. An intelligent, highly motivated person can hide undesirable facets of his or her character in an interview* or on a questionnaire—which together have weeded out applicants with obvious personality disorders—but not so easily under a weeklong observation. In the words of JAXA psychologist Natsuhiko Inoue, "It's difficult to be a good man always." Isolation chambers are also a way to judge things like teamwork, leadership, and conflict management—group skills that can't be assessed in a one-on-one interview. (NASA does not use isolation chambers.)

The observation room is upstairs from the chamber. It is Wednesday, day three of the seven-day isolation. A row of closed-circuit TVs are lined up for the observers, who sit at long tables with their notepads and cups of tea. Three are here now, university psychiatrists and psychologists, staring at the TVs like customers at Best Buy contemplating a purchase. One TV, inexplicably, is broadcasting a daytime talk show.

Inoue sits at the control console, with its camera zooms and microphone controls and a second bank of tiny TV monitors above his head. At forty, he is accomplished for his age and widely respected in the field of space psychology, yet something in his appearance and demeanor makes you want to reach over and pinch his cheek. Like many male employees here, he wears open-toed slippers over socks. As an American, I have large gaps in my understanding of Japanese slipper etiquette, but to me it suggests that JAXA, as much as his house, feels like home. For this week, anyway, it would be understandable; his shift begins at 6 A.M. and ends just after 10 P.M.

---

* As when astronaut Mike Mullane was asked by a NASA psychiatrist what epitaph he'd like to have on his gravestone. Mullane answered, "A loving husband and devoted father," though in reality, he jokes in *Riding Rockets*, "I would have sold my wife and children into slavery for a ride into space."

On camera now, one of the applicants can be seen lifting a stack of 9-by-11-inch envelopes from a cardboard box. Each envelope is labeled with an applicant's identifying letter—*A* through *J*—and contains a sheet of instructions and a square, flat cellophane-wrapped package. Inoue says the materials are for a test of patience and accuracy under pressure. The candidates tear open the packages and pull out sheaves of colored paper squares. "The test is involving . . . I am sorry, I don't know the word in English. A form of paper craft."

"Origami?"

"Origami, yes!" Earlier today, I used the handicapped stall in the hallway bathroom. On the wall was a confusing panel of levers, toggles, pull chains. It was like the cockpit of the Space Shuttle. I yanked a pull-chain, aiming to flush, and set off the emergency Nurse Call alarm. I'm wearing pretty much the same face right now. It's my *Wha?* face. For the next hour and a half, the men and women who vie to become Japan's next astronauts, heroes to their countrymen, will be making paper cranes.

"One thousand cranes." JAXA's chief medical officer, Shoichi Tachibana, introduces himself. He's been standing quietly behind us. Tachibana came up with the test. A Japanese tradition holds that a person who folds a thousand cranes will be granted health and longevity. (The gift is apparently transferable; the cranes, strung on lengths of thread, are typically given to patients in hospitals.) Later, Tachibana will place a perfect yellow crane, hardly bigger than a grasshopper, onto the table where I sit. A tiny dinosaur will appear on the arm of the sofa in the corner. He's like one of those creepy movie villains who sneak into the hero's home and leave behind a tiny origami animal, their creepy villain calling card, just to let him know they were there. Or, you know, a guy who enjoys origami.

The applicants have until Sunday to finish the cranes. Paper squares are spread across the table, the vibrancy of the colors played up by the drabness of the room. Along with the shoebox architecture and the rockets reclining around the grounds, JAXA has managed to duplicate the uniquely unappealing green-gray you often see on NASA interior walls. It's a color I have seen nowhere else and on no paint chip, yet here it is.

The genius of the Thousand Cranes test is that it creates a chronological record of each candidate's work. As they complete their cranes, candidates string them on a single long thread. At the end of the isolation, everyone's string of cranes will be taken away and analyzed. It's forensic origami: As the deadline nears and the pressure increases, do the candidate's creases become sloppy? How do the first ten cranes compare to the last? "Deterioration of accuracy shows impatience under stress," Inoue says.

I have been told that 90 percent of a typical mission on the International Space Station (ISS) is devoted to assembling, repairing, or maintaining the spacecraft itself. It's rote work, much of it done while wearing a pressurized suit with a limited oxygen supply—a ticking clock. Astronaut Lee Morin described his role in installing the midsection of the ISS truss, the backbone to which various laboratory modules are attached. "It's held on with thirty bolts. I personally tightened twelve of them." ("So that's two years of education for each bolt," he couldn't help adding.) The spacesuit systems lab at Johnson Space Center has a glove box that mimics the vacuum of space and inflates a pair of pressurized gloves. In the box with the gloves is one of the heavy-duty carabiners that tether astronauts and their tools to the exterior of the space station while they work. Trying to work the tether is like dealing cards with oven mitts on. Simply closing one's fist tires the hand within minutes. You cannot be the sort of person who gets frustrated easily and turns in a haphazard performance.

An hour passes. One of the psychiatrists has stopped watching and turned his attention to the talk show. A young actor is being interviewed about his wedding and what kind of father he hopes to be. The candidates are bent over the table, working quietly. Applicant A, an orthopedist and aikido enthusiast, is in the lead with fourteen cranes. Most of the rest have managed seven or eight. The instructions are two pages long. My interpreter Sayuri is folding a piece of notebook paper. She is at step 21, where the crane's body is inflated. The directions show a tiny puff beside an arrow pointing at the bird. It makes sense if you already know what to do. Otherwise, it's wonderfully surreal: *Put a cloud inside a bird.*

IT IS DIFFICULT, though delightful, to picture John Glenn or Alan Shepard applying his talents to the ancient art of paper-folding. America's first astronauts were selected by balls and charisma. All seven Mercury astronauts, by requirement, were active or former test pilots. These were men whose nine-to-five involved breaking altitude records and sound barriers while nearly passing out and crashing in screaming-fast fighter jets. Up through Apollo 11, every mission included a major NASA first. First trip to space, first orbit, first spacewalk, first docking maneuver, first lunar landing. Seriously hairy shit was going down on a regular basis.

With each successive mission, space exploration became a little more routine. To the point, incredibly, of boredom. "Funny thing happened on the way to the moon: not much," wrote Apollo 17 astronaut Gene Cernan. "Should have brought some crossword puzzles." The close of the Apollo program marked a shift from exploration to experimentation. Astronauts traveled no farther than the fringes of the Earth's atmosphere to assemble orbiting science labs—Skylab, Spacelab, Mir, ISS. They carried out zero-gravity experiments, launched communications and Defense Department satellites, installed new toilets. "Life on Mir was mostly mundane," says astronaut Norm Thagard in the space history journal *Quest*. "Boredom was the most common problem I had." Mike Mullane summed up his first Space Shuttle mission as "throwing a few toggle switches to release a couple comms satellites." There are still firsts, and NASA proudly lists them, but they don't make headlines. Firsts for shuttle mission STS-110, for instance, include "first time that all of a shuttle crew's spacewalks were based from the station's Quest Airlock." "Capacity to Tolerate Boredom and Low Levels of Stimulation" is one of the recommended attributes on a Space Shuttle–era document drafted by the NASA In-House Working Group on Psychiatric and Psychological Selection of Astronauts.

These days the astronaut job title has been split into two categories. (Three, counting payload specialist, the category into which teachers, boondoggling senators,* and junketing Saudi princes fall.) Pilot astronauts are the ones at the controls. Mission specialist astronauts carry out the science experiments, make the repairs, launch the satellites. They're still the best and the brightest, but not by necessity the boldest. They're doctors, biologists, engineers. Astronauts these days are as likely to be nerds as heroes. (JAXA astronauts on the ISS thus far have been classified as NASA mission specialists. The ISS includes a JAXA-built laboratory module, called Kibo.) The most stressful part of being an astronaut, Tachibana told me, is not getting to be an astronaut—not knowing whether or when you'll get a flight assignment.

---

* Between the astronauts who used their status to win a place in the Senate and the senators who used their influence to win a spot on a NASA mission, there's practically been a Senate quorum in space. (John Glenn managed to work it both ways, returning to space as a seventy-seven-year-old senator.) The gambit occasionally backfires, as when Jeff Bingaman defeated Apollo-astronaut-turned-New-Mexico-senator Harrison Schmitt using the campaign slogan "What on Earth has he done for you lately?"

The first time I spoke to an astronaut, I didn't know about the pilot–mission specialist split. I pictured astronauts, all of them, as they were in the Apollo footage: faceless icons behind gold visors, bounding like antelopes in the moon's weak gravity. The astronaut was Lee Morin. Mission Specialist Morin is a big, soft-spoken man. One foot turns in slightly as he walks. He was dressed in chinos and brown shoes the day we met. There were sailboats and hibiscus flowers on his shirt. He told me a story about how he helped test the lubricant for a launch-pad escape slide on the Space Shuttle. "They had us bend over and they brushed our butts with it. And then we jumped on the slide. And it passed, so [the shuttle mission] could go forward and the space station could be built. I was proud," he deadpanned, " to do my part for the mission."

I remember watching Morin walk away from me, the endearing gait and the butt that got lubed for science, and thinking, "Oh my god, they're just people."

NASA funding has depended in no small part upon the larger-than-life mythology. The imagery forged during Mercury and Apollo remains largely intact. In official NASA 8-by-10 astronaut glossies, many still wear spacesuits, still hold their helmets in their laps, as though at any moment the Johnson Space Center photography studio might inexplicably depressurize. In reality, maybe 1 percent of an astronaut's career takes place in space, and 1 percent of that is done in a pressure suit. Morin was on hand that day as a member of the Cockpit Working Group for the Orion space capsule. He was helping figure out sight lines and optimal placement of computer displays. Between flights, astronauts spend their days in meetings and on committees, speaking at schools and Rotary clubs, evaluating software and hardware, working at Mission Control, and otherwise, as they say, flying a desk.

Not that bravery has been entirely phased out. Those recommended astronaut attributes also include "Ability to Function Despite

Imminent Catastrophe." If something goes wrong, everyone's clarity of mind is needed. Some selection committees—the Canadian Space Agency's, for instance—appear to put greater emphasis on disaster coping skills. Highlights of CSA's 2009 astronaut selection testing were posted in installments on the Web site home page. It was reality television. The candidates were sent to a damage-control training facility, where they learned to escape burning space capsules and sinking helicopters. They leapt feetfirst into swimming pools from terrifying heights while wave generators pushed 5-foot swells. A percussive action-movie soundtrack ramped up the drama. (It is possible the footage had more to do with attracting media coverage than with the realities of choosing Canada's next astronaut.)

Earlier, I asked Tachibana whether he was planning to pull any surprises on his candidates, to see how they cope under the stress of a sudden emergency. He told me he had given thought to disabling the isolation chamber toilet. Again, not the answer I was expecting, but genius in its way. The footage might not play as well with a kettledrum soundtrack (and then again it might), but it's a more apt scenario. A broken toilet is not only more representative of the challenges of space travel, but—as we'll see in chapter 14—stressful in its own right.

"Before you arrived yesterday," Tachibana added, "we delayed lunch by one hour." The little things can be big tells. Unaware that a late lunch or a malfunctioning toilet is part of the test, the applicants behave truer to character. When I first began this book, I applied to be a subject in a simulated Mars mission. I made it past the first round of cuts and was told that someone from the European Space Agency would call me for a phone interview later in the month. The call came at 4:30 A.M., and I did not take care to hide my irritation. I realized later that it had probably been a test, and I had failed it.

NASA uses similar tactics. They'll call an applicant and tell her that they need to redo a couple tests on her physical and that they need to do it the following day. "What they're really doing is saying, 'Let's see if they'll drop everything to be one of us,'" says planetary geologist Ralph Harvey, whose Antarctic Search for Meteorites (ANSMET) program personnel sometimes apply for astronaut openings. (Antarctica is a useful analog for space, and people who thrive there are thought to be psychologically well equipped for the isolation and confinement of space travel.) Harvey recently got a call about one such applicant. "They said, 'We're going to give him a T-38 to fly for the first time tomorrow. And we'd like you to go along with him as an observer and tell us how you think he's doing.' And I said, 'Absolutely.' But I knew that wasn't going to happen. What they were doing was assessing my confidence level in the person."

Another reason to see how would-be astronauts handle stress is that options for reducing it are limited on board a spaceship. "Shopping, let's say," says Tachibana. "You cannot do such a thing." Or drinking. "Or a long bath," adds Kumiko Tanabe, who handles press and publicity for JAXA and thus, I suspect, takes lots of long baths.

LUNCH HAS ARRIVED, and all ten candidates get up to unpack the containers and set out plates. They sit down again, but no one picks up chopsticks. You can tell they're strategizing. Does taking the first bite show leadership, or does it suggest impatience and self-indulgence? Applicant A, the physician, comes up with what seems an ideal solution. "*Bon appétit*," he says to the group. He picks up his chopsticks as the others do, but then waits for someone else to take the first bite. Canny. I've got my money on A.

Here's the other thing that's changed since the heyday of space exploration. Crews aboard space shuttles and orbiting science labs are two or three times the size of Mercury, Gemini, and Apollo crews, and the missions span weeks or months, not days. This makes the Mercury-era "right stuff" the wrong stuff. Astronauts have to be people who play well with others. NASA's recommended astronaut attribute list includes an Ability to Relate to Others with Sensitivity, Regard, and Empathy. Adaptability, Flexibility, Fairness. Sense of Humor. An Ability to Form Stable and Quality Interpersonal Relationships. Today's space agency doesn't want guts and swagger. They want Richard Gere in *Nights in Rodanthe.** Assertiveness has to be "Appropriate" and Risk-Taking Behavior has to be "Healthy." The right stuff is no longer bravado, aggressiveness, and virility. Or as Patricia Santy, NASA's first staff psychiatrist, put it in *Choosing the Right Stuff*, "narcissism, arrogance, and interpersonal insensitivity." "Who," she asks, "would want to work with a person like that?"

As a gross overgeneralization, the Japanese are well suited to life on a space station. They're accustomed to small spaces and limited privacy. They're a lighter, more compact payload than the average American. Perhaps most important, they're raised to be polite and to keep their emotions in check. My interpreter, Sayuri, a woman so considerate she wipes the lipstick off the edge of her teacup before handing it to the JAXA cafeteria dishwashers, says her parents used to tell her, "Don't make waves on the quiet surface of the pond." Being an astronaut, she noted, is "an extension of everyday life." "They make excellent astronauts," agreed Space Shuttle crew member Roger Crouch, whom I had been emailing during my stay in Japan.

I ran my theory by Tachibana. We had gone down to the lobby to chat. We sat on low sofas arranged beneath portraits of the

* It was a ten-hour flight to Tokyo.

JAXA astronaut corps. "What you say is true," he said, one knee bobbing up and down. (His boss told me when I'd visited earlier in the year that leg-bobbing is viewed as a red flag during astronaut selection interviews, along with failure to make eye contact. For the remainder of the conversation, the boss and I stared intently at each other across the table, both refusing to look away.) "We Japanese have a tendency to suppress emotion and try to cooperate, try to adapt, too much. I worry that some of our astronauts behave too much well." Suppressing one's feelings too tightly for too long takes a toll. You either explode or implode. "Most Japanese will become depressive rather than explosive," says Tachibana. Fortunately, he adds, JAXA astronauts train with NASA astronauts for several years, and during those years "their character becomes somewhat more aggressive and like Americans."

In the previous isolation-chamber test, one applicant was eliminated because he expressed too much irritation and another because he was unable to express his irritation and acted it out passively. Tachibana and Inoue look for applicants who manage to achieve a balance. NASA astronaut Peggy Whitson strikes me as a good example. On NASA TV recently, I heard someone at NASA tell her that he could not find a series of photographs that she or some member of her crew had recently taken. If I'd spent the morning shooting photographs and the person I'd shot them for then misplaced them, I'd say, "Look again, lamb chop." Whitson said, without a trace of irritation, "That's not a problem. We can do them over."

Anything else to avoid should you wish to become an astronaut?

Snoring, says Tachibana. If it's loud enough, it can mean elimination from the selection process. "It wakes people up."

According to the *Yangtse Evening Post,* the medical screening for Chinese astronauts excludes candidates with bad breath. Not because it might suggest gum disease, but because, in the words of health screening official Shi Bing Bing, "the bad smell would affect their fellow colleagues in a narrow space."

LUNCH IS OVER, and two—now three, wait, four!—of the candidates are cleaning the surface of the table. I'm reminded of those brushless car washes where a small army of wiping employees descends on your vehicle as it exits the wash. But no one has to clean the dishes. The instructions are to put your dirty plates and utensils back inside the plastic tub labeled with your I.D. letter, and to put the tubs in the "airlock." What the candidates don't know is that the dirty dishes are then loaded onto a dolly and wheeled away to be photographed. The photos will be delivered to the psychiatrists and psychologists, along with the origami birds. I watched the photo shoot after last night's meal. The photographer's assistant opens each tub and holds a piece of cardboard printed with the candidate's letter and the date just inside the bottom of the frame, as though the place setting had been picked up for a crime and was now being posed for a mug shot.

Inoue was vague about the purpose. To see what they ate, he said. For what it's worth, C didn't eat her chicken skin, and G left the seaweed in his miso soup. E left half his soup and all his pickled vegetables. My man A ate everything and placed it back in the container in the same precise configuration in which it had arrived.

"Look at G-san," tutted the photographer. ("San" is a Japanese honorific, like our "Mr." or "Ms.") He lifted the pickle dish that G had placed on top of the dinner plate. "He's hiding his skin."

I'm not sure I understand why it's important that astronauts clean their plates and stack their dirty dishes. Tidiness is certainly important in a small space, but I think this is about something else.

If I showed a stranger a list of the activities I've been observing these past few days and asked him to guess where I'd been, I doubt "space agency" would leap to mind. "Grade school" might. In addition to origami, the tests this week have involved building LEGO robots and making colored-pencil drawings of "Me and My Colleagues" (also destined for the mental health professionals' in-boxes).

Right now, H is on the TV screens, addressing his colleagues and the cameras. The activity is called "self-merits presentation." I had expected something along the lines of a one-way job interview, a recitation of character strengths and job skills. This is more like a summer camp talent show act. C's talent was singing songs in four languages. D did forty push-ups in thirty seconds.

Adding to the overall schoolyard ambiance, the candidates wear pinnies. They're the sort of thing kids used to wear during gym class to help them keep track of who's on what team. These have candidates' letters printed on them. They are for the observers. The lighting is poor and the camera rarely zooms in on faces, so it's hard to figure out who's talking. Before the pinnies went on, everyone was constantly leaning over and whispering to their neighbor. "Who's that? E-san?" "I think it's J-san." "No, J-san is there, with the stripes."

H is saying: "I can ride a bike without holding the handlebars." Now he cups his hands together and puts his lips to his bent thumbs. After a few tries, he produces a low, dry, unmusical whistle. "I don't have a skill like yours," H says to B glumly. B just finished telling us about the badminton championship his team won and then pulling up the legs of his shorts to show off his thigh muscles.

H sits down, and F stands up. F is one of three pilots in the group. "What is important in a pilot is communication." After a solid start, the presentation takes an unexpected turn. F tells us that he often goes out drinking with his pals. "We go to places where ladies entertain. That helps to communicate and help break the ice with the guys." F opens his mouth wide. He's doing something with his tongue. The psychiatrists lean toward the TVs. Sayuri's eyebrows shoot up. "I do this for the ladies," says F. *Wha?* Inoue pulls the zoom. F's tongue is double-curled, like a pair of tacos. "For me it is an ice-breaking technique."

My guy A is up next. He tells us he is going to demonstrate an aikido technique and asks for a volunteer. D stands up. His pinnie is partly slipping off his shoulder like a bra strap. A says that when he was in college, the younger students would get so drunk they couldn't move. "So I twist their arm to help them get up." He grabs D's wrist. D yelps, and everyone laughs.

"They're like frat boys," I say to Sayuri. Tachibana is sitting beside Sayuri, who explains "frat boy" to him.

"To tell you the truth," Tachibana says, "astronaut is a kind of college student." He is given assignments. Decisions are made for him. Going into space is like attending a very small, very elite military boarding school. Instead of sergeants and deans, there is space agency management. It's hard work, and you better stick to the rules. Don't talk about other astronauts. Don't use cuss words.* Never complain. As in the military, wave-makers are leaned on hard or sent away.

All through the space station era, the ideal astronaut has been an exceptionally high-achieving adult who takes direction and follows rules like an exceptionally well-behaved child. Japan cranks

---

* I read an unedited draft of an oral history last week that had the "dangs" and "hells" inked out like operatives in a CIA dossier. When Gene Cernan responded to an Apollo 10 close call with "more than a few goddams, fucks and shits," the president of Miami Bible College wrote to President Nixon demanding public repentance. NASA made Cernan comply. He got the last word in his memoir: "Bunch of goddam hogwash."

them out. This is a culture where almost no one jaywalks or litters. People don't tend to confront authority. My seatmate on the flight to Tokyo told me that her mother had forbidden her to get her ears pierced. It wasn't until she was thirty-seven that she summoned the courage to do it anyway. "I'm just now learning to stand up to her," she confided. She was forty-seven, and her mother was eighty-six.

"Of course, exploration to Mars will be a different story," says Tachibana. "You need someone aggressive, creative. Because they'll have to do everything by themselves." With a twenty-minute radio transmission lag time, you can't rely on advice from ground control in an emergency. "You need again a brave man."

A FEW WEEKS after I left Tokyo, an email arrived from the JAXA Public Affairs Office, informing me that candidates E and G had been selected. E is a pilot with All Nippon Airways and a fan of Japanese musicals. For his self-merits presentation, he acted out a scene from his favorite musical. The scene required E to pretend to weep and wrap his arms around his invisible mother. It was brave, though not in an astronaut sort of way. G is also a pilot—with the Japan Air Self-Defense Force. Military pilots have always been a good fit for the astronaut corps, and not just because of their aviation background and skills. They're used to taking risks and operating under pressure, used to bunking in cramped quarters with no privacy, used to following orders and enduring long separations from their families. Also, as one JAXA staffer pointed out, astronaut selection is political. Air forces have always had ties to space agencies.

The week after I left Japan, all ten candidates flew to Johnson Space Center for interviews with NASA astronauts and selection committee members. Tachibana and Inoue conceded that the applicants' English skills were an important factor in the decision, as was, I imagine, how well they click with the NASA crews. "The most important part of all this, the heart of the process," says ANSMET's Ralph Harvey, "is the interview where they sit you down with a couple astronauts and you just talk. You're someone they may end up stuck in the equivalent of a tent in Antarctica with, for not just six weeks or six months in the space station, but maybe ten years as you're waiting to fly, working at Mission Control or elsewhere. They're picking a buddy as much as they're picking a work partner." A Japanese pilot has an advantage over a doctor in that he has something in common with a lot of NASA astronauts. The military and aviation are global fraternities, and E and G are members.

THE FIRST TIME I visited JAXA, I traveled with a different interpreter. As we drove along the route from the train station, Manami translated some of the signs. One welcomed us to TSUKUBA, CITY OF SCIENCE AND NATURE. I had always heard it called Tsukuba Science City. Not only JAXA is here, but also the Agricultural Research Institutes, the National Institute for Materials Science, the Building Research Institute, the Forestry and Forest Products Institute, the National Institute for Rural Engineering, and the Central Research Institute for Feed and Livestock. There are so many research institutes here that they have their own institute: the Tsukuba Center for Institutes. So what's with the "and Nature" in the city's name? Manami explained that when people first moved to Tsukuba, there weren't any trees or parks or anything to do other than work. No major roads or express trains led into or out of the city. People just worked and worked. There were a lot of suicides, she said, a lot of people jumping off the institute roofs. So the government built a mall and some parks and planted trees and grass, and changed the name to Tsukuba, City of Science and Nature. It seemed to help.

The story made me think about a trip to Mars and what it would be like to spend two years trapped inside sterile, man-made structures with no way to escape one's work and colleagues and no flowers or trees or sex and nothing to look at outside the window but empty space or, at best, reddish dirt. The astronaut's job is stressful for all the same reasons yours or mine is—overwork, lack of sleep, anxiety, other people—but two things compound the usual stresses: the deprivations of the environment and one's inability to escape it. Isolation and confinement are issues of no small concern to space agencies. The Canadian, Russian, European, and U.S. space agencies are spending $15 million on an elaborate psychology experiment that puts six men in a simulated spaceship on a pretend mission to Mars. The hatch opens tomorrow.

2

LIFE IN A BOX

The Perilous Psychology of Isolation and Confinement

Mars is upstairs on the left. The Martian Surface Simulator is one of five locked, interconnected modules that comprise the mission simulation known as Mars500—the number referring to the days needed for a round-trip spin and a four-month stay on Mars. The simulation is taking place on the ground floor of Moscow's Institute of Biomedical Problems (IBMP), Russia's main aerospace medicine research facility. The crew have been paid 15,000 euros each to be subjects in a battery of psychology experiments aimed at understanding and counteracting the baneful effects of being trapped in a small, artificial environment with roommates you did not choose.

Today they "land." Television crews are running up and down the stairs, looking for the best place to plant their tripods. "At first they are all down there," says a bemused IBMP staffer who has been posted on the mezzanine above the Habitable Module. "And now you see the small anthill here."

A recording of military fanfare and some last-minute reportorial elbowing heralds the opening of the hatch. The six men step outside and smile at the cameras. They are accustomed to being filmed. They've been monitored day and night for the past three months. (The shorter isolation served as a practice run for the 500-day simulation scheduled to start in 2010.) The crewmen wave until it begins to seem silly and one by one they drop their arms. They are dressed in blue "flight suits." Walking back to the subway later, I pass the grounds staff of a neighboring apartment complex dressed in the same blue coveralls, bestowing the fleeting impression that cosmonauts are moonlighting as gardeners and handymen.

Isolation-chamber experiments have been a lucrative cottage industry at IBMP for decades. I came across a paper from 1969, detailing a yearlong simulated mission to an unstated destination. The setup was similar to Mars500, though with small, entrancing exceptions, like the "self-massage" that ended each day. The article ran in an academic journal, but you felt as though you were paging through a sort of homosexual *Ladies' Home Journal.* Photographs show the three men preparing dinner, tending plants in the greenhouse, listening to the radio in their turtlenecks and sweater vests, and cutting one another's hair. The journal paper made no mention of spats or maladaptive symptomology, of Bozhko going after Ulybyshev with the barber scissors. The papers rarely include these details. Press conferences don't either. Press conferences are a time for canned speeches and upbeat generalities.

Like this: "We had no problems, no conflicts," Mars500 Commander Sergei Ryazansky is saying. The press conference is being held in a room on the second floor, meaning that most of the camera crews had to fold up their tripods and charge back up the stairwell, affording yet more glee for IBMP staff. There are maybe 200 chairs for 300 bottoms.

"Everyone was supporting each other." After ten minutes of fluff from Ryazansky, a reporter lays it out: "We in the media would like to have some gossip. Can you give some examples of personal tensions?"

They cannot. Pretend astronauts have to be discreet because many of them want to be real astronauts. The Mars500 crew includes one aspiring European astronaut, one aspiring cosmonaut, and two cosmonauts awaiting flight assignments. Volunteering for a simulated mission is a way to show the space agencies you've got at least some of what it takes: A willingness to adapt to a situation, rather than trying to change it. Tolerance for confinement and stripped-down living conditions. Emotional stability. An accommodating family.

Another reason Ryazansky won't gossip about his crewmates is that, like most isolation chamber volunteers, he signed a confidentiality agreement. Space agencies want to know what happens when you lock people in a box with no privacy and not enough sleep and depressing food, but they are wary of letting the rest of us know. "If a space agency comes out and says, 'Oh, all of these problems happen,' then people say, 'Oh, all of these problems happen! Why do we go to space? It's too risky!'" says Norbert Kraft, a physician who now researches group psychology and productivity on long-duration missions for NASA's Ames Research Center in California. "The agencies try to keep the best image up, otherwise they don't get funded anymore." What happens in the Habitable Module stays in the Habitable Module.

Unless someone blabs, as happened the last time IBMP hosted an isolation. SFINCSS (Simulated Flight of International Crew on Space Station) made minor headlines in 1999 when stories of drunken brawling and sexual assault were leaked to the press. The current crew has obviously been coached for discretion.

"Our personal training allowed us to avoid any conflicts," Ryazansky continues. "Reactions to emotions were really respectful and really, really polite." All around the room, journalists begin to realize they've traveled hundreds of miles for a nonstory. Soon there are enough chairs for everyone.

The SFINCSS "incidents" took place three months into the isolation, when crews in separate modules "docked." One crew consisted of four Russians; the other was (intentionally) a cross-cultural grab bag: a Canadian woman, a Japanese man, a Russian man, and their commander, Austrian-born Norbert Kraft. At 2:30 A.M. on New Year's Day, 2000, the Russian crew commander, Vasily Lukyanyuk, pushed Canadian crew member Judith Lapierre out of range of the cameras and French-kissed her twice, against her protestations. Shortly before the kissing incident, two other Russian subjects got into a fistfight that left the walls spattered with blood. In the aftermath, the hatch between the two modules was shut, the Japanese crew member quit, and Lapierre complained to IBMP and to the Canadian Space Agency. IBMP psychologists, she says, were unsupportive, accusing her of overreacting. Despite having signed a confidentiality agreement and aspiring to become an astronaut, Lapierre told her story to the press. To quote IBMP psychologist Valery Gushin, she "washed her dirty clothes in public."

By the time I contacted Lapierre, she was done with her laundry. She confirmed the basic facts and referred me to her SFINCSS commander, Norbert Kraft. Kraft has spent time on both sides of the closed-circuit TVs—as a consultant on an isolation test at the Japanese Aerospace Exploration Agency and doing time in SFINCSS. He volunteered, he said, out of a desire to know what it's like for the subjects he monitors. Kraft possesses a delightful, free-range curiosity. His SFINCSS bio states that he enjoys waltzing, scuba-diving, cooking black cherry cake, and tending a Japanese stone garden. He was happy to drive all the way up to Oakland from Mountain View to talk with me, because, he said, "it's something different."

Kraft's portrayal of the events was more nuanced than those

in the newspapers. Lapierre was less a victim of sexual harassment than of institutional sexism. To paraphrase Gushin, Russian men prefer that women act like women, not equals—even if they're astronauts. According to Soviet/Russian space program historian Peter Pesavento, U.S. astronaut Helen Sherman was criticized by her crewmates on Mir for what was perceived as an overly professional demeanor—i.e., she didn't flirt. In the decades after Valentina Tereshkova snagged the "First Woman in Space" title for the Soviet Union, in 1963, only two women have flown as cosmonauts. The first, Svetlana Savitskaya, was handed a floral-print apron when she floated through the Salyut hatch.

From the beginning, the IBMP staff and psychologists had been dismissive of Lapierre. They didn't take her seriously as a researcher, because, Kraft says, she's a woman. Not helping: language barriers. Lapierre spoke little Russian and "ground control" spoke little English.* Inside the Russian module, only the commander could converse easily in English. He was kind to Lapierre, and Kraft believes she saw him as a potential ally in her efforts to gain the Russians' respect. Thus she did what she could to foster the bond. She was friendly, says Kraft, in a way that Russian women are usually not: sitting on his lap, kissing him on the cheek. "She was sending the wrong signal, but she didn't see it."

Kraft says Lapierre was unjustly blamed for the Japanese participant quitting. The man, Masataka Umeda, claimed to have acted out of solidarity with Lapierre. Kraft says Umeda shut the

---

* A common theme throughout Russian-American space collaborations. NASA psychologist Al Holland tells the tale of driving a carful of Russians across Moscow during the Shuttle-Mir program. When the cars in his lane slowed to a standstill, a Russian man in the back seat asked, "What's going on up there?" Holland was proud to be able to use a new vocabulary word: *stopka*—traffic jam. Only he used *popka*: "It's a big rear end!"

hatch because he was bothered by the Russian crew watching porn and that he had been looking for an excuse to bail.

I might have looked for one too. Along with the considerable stress of confinement, sleep deprivation, language and cultural gaps, and lack of privacy, more subtle torments plagued the crew. The shower room had cockroaches and no hot water. Night after night, dinner was kasha ("wheat gruel," Lapierre called it). "Mice came through the floor and mold crawled up the conduits," said Kraft in an email that included six photographs, one with the caption "Hairlice." The lice outbreak didn't bother Kraft—"It's something new"—and the Russian crew calmly shaved their heads. Lapierre had to cope not only with the stress of lice, but with the IBMP staff's response. "The Russians said, 'Judy got a package from Canada that included the lice,'" Kraft recalls.

As producers of reality television know, there is no more reliable way to ignite smoldering frustrations than to douse them in alcohol. On the record, there was only one bottle of champagne, provided by IBMP for the 2000 Millennium Eve. In reality, there were many bottles, not just champagne, but vodka and cognac. Kraft says they find their way into isolation chambers as bribes. If you want the Russian volunteers to do a good job with your research, he says, you "better pack vodka and a salami with your experiment."

Apparently this was also the case on Soviet and Russian space labs. Mir astronaut Jerry Linenger writes in his memoir that he was surprised to find a bottle of cognac in one arm of his spacesuit and a bottle of whiskey in the other. (Linenger was the Frank Burns of space exploration: "I complied strictly with the NASA policy of no alcohol consumption on duty.") On long Russian missions, Kraft says, "You better hide the disinfectant." While I was in Russia, a cosmonaut, who requested anonymity, showed me one of his slides from space: two crew members with straws, floating on either side of a 5-liter tank of cognac like teenagers sharing a malt.

Though the press coverage of SFINCSS put IBMP and the space agencies on the defensive, the researchers were pleased to be, as JAXA psychologist Natsuhiko Inoue put it, "getting very unique results." This was, after all, a study of group interactions on cross-cultural missions. "The incident," Inoue told me in an email, "brought us very many valuable insights on future crew selection and training." Mostly commonsense stuff. Make sure they speak a common language well enough to communicate. Check out how well they work as a team. Choose people with a resilient sense of humor. Give everyone a crash course in cross-cultural etiquette. Someone should have warned Lapierre, for example, that "it's nothing" (Gushin's words) for a Russian man to kiss a woman at a party. And that if you want him to stop, you slap him. That "no" means "maybe." And that when Russian men bloody each other's noses, it's "a friendly fight." (Kraft confirmed this surprising item. "It's how they settle disputes. They did it on Mir.")

No matter how thoroughly you try to anticipate cross-cultural clashes, something's bound to be overlooked. Ralph Harvey, who oversees teams of meteorite hunters at remote field camps in Antarctica, told me about a Spanish team member with a habit of plucking hairs from his head and holding them in the flame of the camp stove. "In Spain," the man explained, "the barbers burn the tips of your hair, and I like the smell." For the first week, his tentmate was amused, but it soon became a source of friction. "It's on the questionnaire now," joked Harvey. "*Do you burn your own hair for fun?*"

Kraft believes the media coverage of SFINCSS was beneficial in that it provided a rare honest portrayal of the emotions that develop among men and women confined together in space. He takes issue with the way space agencies portray astronauts as superhuman. "As if they don't have any hormones, they don't have any feelings for anybody." It comes back yet again to a fear of negative publicity and diminished funding. The danger is that an organization invested in downplaying psychological problems is unlikely to spend much time investigating solutions to those problems. "Until," as Kraft puts it, "one of the astronauts goes with diapers* across the U.S. Now they are people suddenly!" (Two days after astronaut Lisa Nowak's infamous confrontation with love rival Colleen Shipman, NASA ordered a review of its psychological screening and evaluation processes for astronauts.)

Making things worse: Astronauts themselves try to hide emotional problems, out of fear they'll be grounded. Access to psychologists is available during missions, but crew are reluctant to make use of it. "Every communication to them means a special notice in your flying book," cosmonaut Alexandr Laveikin told me. "So we were always trying not to ask for specialists' help." Laveikin's Mir mission with Yuri Romanenko was mentioned in a *Quest* article by Peter Pesavento on the psychological effects of space travel. Pesavento says Laveikin returned early from the mission due to "interpersonal issues and cardiac irregularity." (I was to meet with Laveikin and Romanenko the next day.)

It's a dangerous state of affairs. If someone on board a spacecraft is reaching the breaking point, it's important for ground control to know about it. People's lives depend on them knowing that.

---

* Did she or didn't she? Arresting officer William Becton wrote in his affidavit that he found a trash bag containing two used diapers inside Lisa Nowak's car. "I asked Mrs. Nowak why she had the diapers. Mrs. Nowak said that she did not want to stop and use the restroom, so she used the diapers to collect her urine." That's what astronauts do—you can't take a bathroom break on a spacewalk, so you wear a diaper inside your spacesuit.

Nowak later denied wearing diapers. She now says her family had used the diapers when Houston was evacuated during Hurricane Rita, two years earlier. If I were Nowak, I wouldn't have worried about the diapers. I'd have worried about the buck knife, steel mallet, BB gun, gloves, rubber tubing, and large plastic garbage bags also found in her car. I'd be peeing my pants.

This perhaps explains why so many space psychology experiments these days focus on ways to detect stress or depression in a person who doesn't intend to tell you about it. If technologies being tested on Mars500 pan out, spacecraft—and other high-stress, high-risk workplaces like air-traffic control towers—will be outfitted with microphones and cameras hooked up to automated optical and speech-monitoring technology. The robotic spies can detect telltale changes in facial expressions or speech patterns and, hopefully, help those in command to avert a crisis.

The stigma of psychological problems also makes them difficult to study. Astronauts are reluctant to sign on as study subjects, lest the researchers uncover something unflattering. The last time I spoke to NASA consulting psychologist Pam Baskins, she was about to begin an experiment comparing different sleep medications and dosages. The astronauts were to be woken from a sound sleep to see how the drugs affected their ability to function in a simulated middle-of-the-night emergency. It appealed to my sense of fun, and I asked if I could come watch. "Absolutely not," replied Baskins. "It took me a *year* to convince these guys to participate."

A SPACE STATION is a rangy monstrosity, a giant Erector Set assembled by a madman. But the living area inside the Mir core module, where cosmonauts Alexandr Laveikin and Yuri Romanenko spent six months together, would fit in a Greyhound bus. The sleep chambers are less like bedrooms than like phone booths. They have no doors. My interpreter Lena and I are inside a mock-up of the module, in the Memorial Museum of Cosmonautics, in Moscow. With us is Laveikin, who now runs the museum. Yuri Romanenko is on his way. I thought it would be interesting to talk with them inside the room that nearly drove them mad.

Laveikin looks little changed from his official portrait, where he conveys an impression of guileless good cheer. He kisses our hands as though we're royalty. It's neither affectation nor flirtation, just something that Russian men of his era were taught to do. He wears beige linen pants, a splash of cologne, and the cream-colored summer footwear I've been seeing all week on the feet of the men across from me in the Metro.

Laveikin waves hello to a narrow-girdled, suntanned man in jeans, with sunglasses hooked in the V of his shirt collar. It's Romanenko. He is cordial, but not a hand-kisser. Cigarette smoke has roughed up his vocal cords. The two embrace. I count the seconds. *One Mississippi, two Mississippi, three.* Whatever happened between them, it's forgotten or forgiven.

Sitting inside the mock-up, it is easy to imagine how a room this size, for that long, could set two men against each other. Romanenko points out that enclosed spaces are not a necessary ingredient for feeling trapped with someone. "Siberia is a big, big space here in Russia. But our hunters who go to *taiga* [forest] for half year, they're trying to go on their own, just with a dog." Romanenko sits where he used to sit on Mir, in the left-hand spot at the control console, on a backless seat with a bar for hooking one's feet. (Later space stations dispensed with seats, because zero gravity dispenses with sitting.) "Because if there are two or three of you go, it will be conflict."

"And this way," Laveikin grins, "you can eat the dog at the end."

Psychologists use the term "irrational antagonism" to describe what happens between people isolated together for more than about six weeks. A 1961 *Aerospace Medicine* paper included a fine example, from the diary of a French anthropologist who spent four months in the Arctic with a Hudson's Bay fur trader:

> I liked Gibson as soon as I saw him. . . . He was a man of poise and order, he took life calmly and philosophically. . . .

> But as winter closed in around us, and week after week our world narrowed until it was reduced to the dimensions of a trap . . . I began to rage inwardly and the very traits . . . which in the beginning had struck me as admirable, ultimately seemed to me detestable. The time came when I could no longer bear the sight of this man who was unfailingly kind to me. That calm which I had once admired I now called laziness, that philosophic imperturbability became in my eyes insensitiveness. The meticulous organization of his existence was maniacal old-manliness. I could have murdered him.

Likewise, Admiral Richard Byrd preferred to carry out his winter-long weather observations in Antarctica by himself, in perilous conditions and twenty-four-hour darkness, rather than face, as he put it in *Alone*, the moment when "one has nothing left to reveal to the other, when even his unformed thoughts can be anticipated, his pet ideas become a meaningless drool, and the way he blows out a pressure lamp or drops his boots on the floor or eats his food becomes a rasping annoyance."

Other people are just one of the psychological hardships that space serves up. Norbert Kraft summed it up nicely. I had asked him if he thought being an astronaut was the best or the worst job in the world. "You're sleep-deprived, and you have to perform perfectly or else you don't fly anymore. As soon as you're done with something, ground control is telling you something else to do. The bathroom stinks, and you have noise all the time. You can't open a window. You can't go home, you can't be with your family, you can't relax. And you're not well paid. Can you get a worse job than that?"

Laveikin says his 1987 stint on Mir was a hundred times harder than what he had expected. "It's hard work, dirty work. Very noisy, very hot." He had motion sickness for more than a week and no drugs to help him through it. He recalls turning to his commander during the first few days, saying, "Yuri. And we will stay here for *half year*?" To which Romanenko, using Laveikin's nickname, replied, "Sasha, but people stay in prisons for ten years or more."

The bottom line is that space is a frustrating, ungiving environment, and you are trapped in it. If you're trapped long enough, frustration metastasizes to anger. Anger wants an outlet and a victim. An astronaut has three from which to choose: a crewmate, Mission Control, and himself. Astronauts try not to vent at each other because it makes a bad situation worse. There's no front door to slam or driveway to speed out of. You're soaking in it. "Also," says Jim Lovell, who spent two weeks on a loveseat with Frank Borman during Gemini VII, "you're in a risky business and you depend on each other to stay alive. So you don't antagonize the other guy."

Laveikin and Romanenko say they managed to avoid frictions because of the clear hierarchy afforded by age and rank. "Yuri is older than me and had experience of spaceflight," Laveikin is saying. "So naturally he was the leader, the psychological leader. I was following him. And I accepted this role. Our flight was calm."

This is difficult to believe. "You never got mad?"

"Of course," says Romanenko. "But mainly it was flight control center's fault." Romanenko went with option 2. Venting your frustration at Mission Control personnel is a time-honored astronaut tradition, known in psychology circles as "displacement." Sometime around the sixth week of a mission, says University of California, San Francisco, space psychiatrist Nick Kanas, astronauts begin to withdraw from their crewmates, become territorial, and displace their hostility for each other onto Mission Control.

Jim Lovell seemed to do most of his displacing on the Gemini VII nutritionist. "Note to Dr. Chance," says Lovell to Mission Control at one point in the mission transcript. "It looks like we're in a snow storm with crumbs from the beef sandwiches. At 300 dollars a meal! I think you can do better than this." Seven hours later, he gets back on the mic: "Another memo to Dr. Chance: Chicken with vegetables, Serial Number FC680, neck is almost sealed shut. You can't even squeeze it out. . . . Continuing same memo to Dr. Chance: Just opened the seals; chicken with vegetables all over window at this time."

Lovell's mission was only two weeks long. Was the capsule's tiny size accelerating the effects of confinement? Kanas knew of no formal studies, but he confirmed that the smaller the craft, generally speaking, the tenser the astronauts.

Displacement perhaps explains why Judith Lapierre's anger was directed more at IBMP and the Canadian Space Agency than at the Russian commander, whose actions she put down to cross-cultural misunderstanding and "natural man-woman situations." Though it's also easy to believe she directed her anger toward IBMP because they were being *popkas*.

Romanenko retains some residual steam to this day. "People who prepared tasks for us, they have no idea what on board is like. Say you are running something here"—he turns to indicate the Mir control console—"and somebody gives you an order to switch on something else. They don't understand it's over on the other side, and I can't leave what I do here and go there." (This is why space agencies tend to use astronauts as "cap coms"—capsule communicators.) According to Robert Zimmerman's history of the Soviet space stations, Romanenko had, by the final stages of the mission (after Laveikin left), grown so "testy" with the flight control center that his crewmates took over all communications with the ground.

Alexandr Laveikin took the third option. He turned the hostility inward. The result, familiar to any psychologist who deals with isolated, confined populations, is depression. Later, after Romanenko leaves, Laveikin confides that there were moments when he thought about suicide. "I wanted to hang myself. Of course, it's impossible because of weightlessness."

Romanenko predicts trouble on a Mars mission. "*Five hundred days,*" he says with evident horror. Romanenko remained for another four months after Laveikin left. Zimmerman writes that he became increasingly unstable and uncooperative, "devoting his time to writing poems and songs" and exercising. I ask Lena to ask him about this phase of the mission. Earlier, I had told her I'd like to hear some of the songs Romanenko composed in space, and this is what she asks about.

"You want us to sing?" Romanenko laughs his grainy laugh. "We would need fifty grams of whiskey!" I apologize for not having brought any.

"I have it," Laveikin says. "In my office."

It's 11 A.M. But I am not Jerry Linenger.

Laveikin leads us through the museum, narrating as he walks. Here are the giants of Soviet rocketry, one per glass display case. Earlier today, I visited a Moscow natural history museum, and sections of it were arranged in this way—not by taxonomy or ecological niche, but by guy: field notebooks from expeditions, some prized specimens, honors from the tsar. The rocket engineers are represented largely by accessories: pens and wristwatches, eyeglasses and flasks.

In his office, Laveikin sits down to look on his computer for a recording of a song Romanenko wrote while on board Mir. The surface of his desk is mostly empty. An appendage like a gangplank protrudes from the front of it. Laveikin gets up to unlock a liquor cabinet and sets down a bottle of Grant's whiskey and four crystal shot glasses on the plank. It's a bar. In Russia you can buy a desk with a built-in bar!

Laveikin raises his glass. "To . . ." He searches for the words in English. "A nice psychological situation!"

We clink our glasses and empty them. Laveikin refills them. Romanenko's song is playing, and Lena translates: "Sorry Earth, we say good-bye to you . . . our ship is going upwards. . . . But the time will come when we will drop into the blueness of the dawn, as a morning star." And the chorus: "I will fall into the grass and fill my lungs with air. I will drink water from the river. . . ." It's a catchy pop tune, and I'm bopping in my seat until I notice that the lyrics are making Lena sad. "I will kiss the ground, I will hug my friends. . . ." Lena wipes a tear as the song ends.

People can't anticipate how much they'll miss the natural world until they are deprived of it. I have read about submarine crewmen who haunt the sonar room, listening to whale songs and colonies of snapping shrimp. Submarine captains dispense "periscope liberty"—a chance to gaze at clouds and birds and coastlines* and remind themselves that the natural world still exists. I once met a man who told me that after landing in Christchurch, New Zealand, after a winter at the South Pole research station, he and his companions spent a couple days just wandering around staring in awe at flowers and trees. At one point, one of them spotted a woman pushing a stroller. "*A baby!*" he shouted, and they all rushed across the street to see. The woman turned the stroller and ran.

Nothing tops space as a barren, unnatural environment. Astronauts who had no prior interest in gardening spend hours tending experimental greenhouses. "They are our love," said cos-

---

* And to keep their distance vision from deteriorating. When your view extends no farther than a few yards, the muscles that squeeze the lens for near focus can eventually lock in a short-lived "accommodative spasm." Submarine myopia is enough of a problem that submarine crews aren't allowed to drive for from one to three days after coming ashore from a long assignment—a good idea for several reasons.

monaut Vladislav Volkov of the tiny flax plants* with which they shared the confines of Salyut 1, the first Soviet space station. At least in orbit, you can look out the window and see the natural world below. On a Mars mission, once astronauts lose sight of Earth, there'll be nothing to see outside the window. "You'll be bathed in permanent sunlight, so you won't even see any stars," astronaut Andy Thomas explained to me. "All you'll see is black."

Humans don't belong in space. Everything about us evolved for life on Earth. Weightlessness is an exhilarating novelty, but floaters soon begin to dream of walking. Earlier Laveikin told us, "Only in space do you understand what incredible happiness it is just to walk. To walk on Earth."

Romanenko missed the smells of Earth. "Can you imagine being even one week in a locked car? Smell of metal. Smell of paint, rubber. When girls were writing us letters, they were putting drops of French perfume on there. We loved those letters. If you smell a letter from a girl before you go to bed, you see good dreams." Romanenko finishes his whiskey and excuses himself. He hugs Laveikin again and shakes our hands.

---

* If the plants are edible, a conflict can arise. As much as astronauts miss nature, they miss fresh food. The diary of cosmonaut Valentin Lebedev includes a story about a batch of onion bulbs taken on board Salyut as part of an investigation of plant growth in zero gravity. "As we were unloading the resupply ship, we found some rye-bread and a knife. So we ate some bread. Then we saw the onion bulbs we were supposed to plant. We ate them right then and there, with bread and salt. They were delicious. Time went by and the biologists asked us, 'How are the onions?'

"'They are growing,' we answered. . . .

"'Do they have shoots?' Without any hesitation we replied that they even had shoots. There was great excitement at the communication station. Onions have never bloomed in space before! We asked to speak to the head biologist in private. 'For god's sake,' we told him, 'don't get upset, we ate your onions.'"

I'm trying to imagine NASA filling resupply vehicles with sacks of love letters. Laveikin says it's true. "From all over the Soviet Union, girls were writing letters."

"To girls," I say. Glasses are raised.

"You really feel the absence of a woman," Laveikin tells us. With Romanenko gone, he speaks more freely. "There are sexual dreams, as a substitute. It's constant through the flight. We were even discussing that maybe we have to take something from the sex shops. It was discussed at IBMP."

I turn to Lena. What does he mean? "An artificial vagina?"

"*Vagine?*" asks Lena. A discussion ensues. Lena turns back to me. "A mock-up."

Laveikin breaks into English, as he does sometimes to tweak a translation: "A rubber woman." A blow-up doll. Ground control, he says, nixed the idea. "They said, 'If you would do that, then we would need to put it in your schedule for the day.'

"We have a joke. You know we have food in tubes." I do. Tubes of space borscht are on sale in the museum gift shop. "There are white and black tubes. On the white is written BLONDE. On black one: BRUNETTE.

"But please understand, sexual concerns are far from being the dominant concerns in space. It's down here on the list." With his hand, he indicates a level down by his knee. "It would just be a nice supplement. But when we talk about five hundred days, it's true, this problem starts to grow higher on the list." He believes a Mars crew should be made up of couples, to help ease the tension that builds during a long mission. According to Norbert Kraft, NASA has considered sending married couples into space. When they asked his opinion on the matter, he discouraged it. His reasoning was that an astronaut might find himself with an untenable choice: jeopardizing his spouse or jeopardizing the mission. Astronaut Andy Thomas, who is married to astronaut Shannon Walker, told me another reason NASA shies away from flying married couples. In the event of a crash or explosion, they don't want one family to have to endure a double loss, particularly if the couple has children.

Laveikin listens, then amends his statement: "Not necessarily married."

"That's right," says Lena. "There would be a different ethic there. When you come back to Earth, your wife should understand that at that time it was like different dimension, different rules, different you."

Laveikin laughs. "My wife is a clever person. She would understand. She'd say, 'You're not completely faithful even on Earth. Let it be in space as well.'"

Kraft would agree. He told me he advocates sending nonmonogamous couples—straight and/or gay—to Mars. "[Space agencies] are going to have to be more liberal and open about that. Mix and match or whatever." Andy Thomas imagines that happening naturally on a Mars mission—as it tends to in Antarctica. "It's very common for people there to pair off and form sexual relationships that last through the duration of their stay—to gravitate to a support structure to help them get through the experience. And then at the end of the season, it's all over."

For seventeen years, only men worked the research bases in Antarctica. Women, the excuses went, mean trouble: distraction, promiscuity, jealousy. It wasn't until 1974 that the McMurdo Station winter-over personnel included women. One was a spinster biologist in her fifties who appears in photographs wearing a gold cross over her turtleneck. The other was a nun.

These days, a third of U.S. Antarctic personnel are women. They are credited with a rise in productivity and emotional stability. Mixed-gender crews are, as Ralph Harvey puts it, more "middle-of-the-bell curve." There are fewer fistfights and fart jokes. "No one hurts his back lifting too big of a box." Norbert Kraft told me about a teamwork study he ran at NASA Ames that compared all-male, all-female, and mixed-gender teams. The mixed-gender groups performed best. (The lowest scores belonged to the all-woman teams. "You can't have all the chitchatting," Kraft said bravely.)

Laveikin: "Can you imagine six men on the way to Mars, what will happen?"

"I know," I say, though I'm not entirely sure we're imagining the same thing. "Look what happens in prisons."

"And on submarines. And geologists in the field."

I make a note to ask Ralph Harvey about this. Laveikin quickly adds that he cannot recall hearing of any instances of "man-on-man love" in the Russian cosmonaut corps.* In the end, the least problematic Mars crew might be the kind Apollo astronaut Michael Collins (jokingly) suggests in his memoir: a "cadre of eunuchs."

THE FIRST AEROSPACE isolation chambers held just one man. The Mercury and Vostok psychiatrists didn't worry about crew members getting along with one another; the flights were a few hours or, at most, a couple days long, and the astronauts flew solo. What the psychiatrists worried about was space itself. What happens to a man alone in a silent, black, endless vacuum? To find out, they tried to approximate space here on Earth. Researchers at the Aeromedical Research Laboratory at Wright-Patterson Air Force Base

---

* Yuri Gagarin loved Soviet rocketry mastermind Sergei Korolev, though not in a space food tube sort of way. When searchers found Gagarin's wallet after the fighter jet crash that killed him, there was just one photo inside (now on display beside the mangled wallet in the Star City museum). The photo is of Korolev—not Gagarin's wife or child, not his beloved mother. Not even Gina Lollobrigida. "She kissed him!" said our ebullient museum guide Elena while fanning herself with a plastic fan as though overcome by the thought of it.

soundproofed a 6-by-10-foot commercial walk-in freezer, put a cot, some snacks, and an enamel chamber pot inside, and turned off the lights. A three-hour stint in the isolation chamber became one of the Mercury astronaut qualifying tests. One account I read, by a Mercury aspirant named Ruth Nichols, described it as the toughest test the candidates endured. Some male pilots, Nichols said, "responded violently" after only a few hours.

Colonel Dan Fulgham was in charge of the Wright-Patterson tests. He doesn't recall any Mercury candidates becoming violent or otherwise "losing it" during their isolation test. He recalls them using it to catch up on sleep.

The researchers soon began to realize that sensory deprivation was a poor approximation of spaceflight. Space is black, but there's plenty of sunlight, and the capsules would be lighted. Radio contact would be possible much of the time. Claustrophobia and solitude were the more salient concerns, especially on a longer mission. That is why, in 1958, an airman from the Bronx, named Donald Farrell, undertook a two-week pretend moon mission in the One-Man Space Cabin Simulator at the School of Aviation Medicine, at Brooks Air Force Base in Texas. A *Time* magazine article described his (sadly long-lost) diary as being increasingly obscenity-laden, but in newspaper interviews he complained only that he missed cigarettes and forgot his comb. Farrell's greatest hardship, by my reckoning, was the recording of "Love Is a Many-Splendored Thing" and other "soft music" piped into the simulator.

In retrospect, it was silly to think that the experience of traveling in space could be approximated by a repurposed walk-in freezer.

To find out what would happen to a man alone in the cosmos, at some point you just had to lob one up there.

3
STAR CRAZY

Can Space Blow Your Mind?

Yuri Gagarin stands on a pedestal two stories high, in a patch of grass by a Moscow thoroughfare. You can tell at a distance that it's him by the way his arms are posed—out away from his sides, fingers pressed together, in the manner of the flying superhero. From the base of the monument, looking up, you cannot see the head of the first man in space, just the heroic chest and the tip of the nose protruding beyond it. I'm joined by a man in a black shirt with a bottle of Pepsi under one arm. His head is lowered, which I take to be a gesture of respect until I see that he is clipping his fingernails.

Nationalist glory aside, Gagarin's 1961 flight was primarily a psychological achievement. His task was simple, though by no means easy: Climb inside this capsule and let us blast you, alone and at great peril, past the borderline of space. Let us catapult you into an airless, lethal nothingness, where no man has ever been. Whip around the planet, and then come down and tell us what it was like for you.

There was a great deal of conjecture at the time—both at the Soviet space agency and at NASA—about the unique psychological consequences of breaching the cosmos. Would hurtling into "the black," as pilots used to call it, blow the astronaut's mind? Hear the ominous words of psychiatrist Eugene Brody, speaking at the 1959 Symposium on Space Psychiatry: "Separation from the earth with all of its unconscious symbolic significance for man, . . . might in theory at least be expected . . . to produce—even in a well-selected and trained pilot—something akin to the panic of schizophrenia."

There was worry that Gagarin might come unhinged and sabotage the history-making mission. It was enough of a worry that the powers-that-be locked the manual controls of the Vostok capsule before liftoff. What if something went awry and communications went dead and Pilot-Cosmonaut #1 needed to take control of the capsule? His superiors had thought about that too, and seemingly turned to game show hosts for advice. Gagarin was given a *sealed envelope* containing the secret combination to unlock the controls.

The concerns were not altogether fatuous. In a study published in the April 1957 issue of *Aviation Medicine*, 35 percent of 137 pilots interviewed reported having experienced a strange feeling of detachment from Earth while flying at high altitudes, almost always during a solo flight. "I feel like I have broken the bonds from the terrestrial sphere," said one pilot. The phenomenon was pervasive enough for psychologists to give it a name: the breakaway effect. For a majority of these pilots, the feeling wasn't one of panic, but of euphoria. Only 18 of the 137 characterized their feelings as fear or anxiety. "It seems so peaceful, it seems like you are in another world." "I feel like a giant." "A king," said another. Three commented that they felt nearer to God. A pilot named Mal Ross, who set a series of altitude records in experimental aircraft in the late 1950s, twice reported an eerie "feeling of exultation, of wanting to fly on and on."

The year the *Aviation Medicine* article ran, Colonel Joe Kittinger ascended to 96,000 feet in an upright, phone-booth-sized sealed capsule suspended beneath a helium balloon. With his oxygen dangerously low, Kittinger was ordered by his superior, David Simons, to begin his descent. "COME AND GET ME," replied Kittinger, letter by letter in Morse code. Kittinger says it was a joke, but Simons didn't take it that way. (Morse code has always been a tough medium for humor.) In his memoir *Man High,* Simons recalls thinking that "the weird and little understood breakaway phenomenon could be taking hold of Kittinger's mind, . . . that he . . . was gripped in this strange reverie and was hellbent on flying on and on without regard for the consequences."

Simons compared the breakaway phenomenon to "the deadly raptures of the deep." "Rapture of the deep" is a medical condition—a feeling of calm and invulnerability that can steal over a diver, usually at depths below 100 feet. It is more prosaically known as nitrogen narcosis, or as the Martini Effect (one drink for every 33 feet below 65 feet). Simons speculated that one day soon aerospace physicians would be talking about a condition "known as the deadly rapture of space."*

---

* Every mode of travel has its signature mental aberration. Eskimo hunters traveling alone on still, glassy waters are sometimes stricken by "kayak angst"—delusions that their boat is flooding or that the front end is either sinking or rising up out of the water. Of related interest: "A Preliminary Report of Kayak-Angst Among the Eskimo of West Greenland" includes a discussion of Eskimo suicide motives and notes that four out of the fifty suicides investigated were elderly Eskimos who "took their lives as a direct result of uselessness due to old age." No mention was made of whether they cast themselves adrift on ice floes, as you sometimes hear, and whether travel by ice floe has its own unique anxiety syndrome.

He was right, though NASA preferred the less flowery term "space euphoria." "Some NASA shrinks," wrote astronaut Gene Cernan in his memoir, "had warned that when I looked down and saw the Earth speeding past so far below, I might be swamped by space euphoria." Cernan would soon be undertaking a spacewalk—history's third—during Gemini IX. The psychologists were nervous because the first two spacewalkers had expressed not only an odd euphoria but a worrisome disinclination to go back inside the capsule. "I felt excellent and in a cheerful mood and reluctant to leave free space," wrote Alexei Leonov, the first human to, in 1965, float freely in the vacuum of space, attached to his Vokshod capsule by an air hose. "As for the so-called psychological barrier that was supposed to be insurmountable by man preparing to confront the cosmic abyss alone, I not only did not sense any barrier, but even forgot that there could be one."

Four minutes into NASA's first spacewalk, Gemini IV astronaut Ed White gushed that he felt "like a million dollars." He struggled to find the words for it. "I've . . . it's just tremendous." There are moments when the mission transcript reads like the transcript of a 1970s encounter group. Here are White and his commander, James McDivitt, a couple of Air Force guys, after it's over:

WHITE: That was the most natural feeling, Jim.
McDIVITT: . . . You looked like you were in your mother's womb.

NASA's concern was not that their astronaut was euphoric, but that euphoria might have overtaken good sense. During White's twenty minutes of bliss, Mission Control repeatedly tries to break in. Finally the capsule communicator, Gus Grissom, gets through to McDivitt.

GRISSOM: Gemini 4, get back in!

McDIVITT: They want you to come back in now.

WHITE: Back in?

McDIVITT: Back in.

GRISSOM: Roger, we've been trying to talk to you for awhile here.

WHITE: Aw, Cape, let me just [take] a few pictures.

McDIVITT: No, back in. Come on.

WHITE: . . . Listen, you could almost not drag me in, but I'm coming.

But he wasn't. Two more minutes passed. McDivitt starts to plead.

McDIVITT: Just come on in . . .

WHITE: Actually, I'm trying to get a better picture.

McDIVITT: No, come on in.

WHITE: I'm trying to get a picture of the spacecraft now.

McDIVITT: Ed, come on in here!

Another minute passes before White makes a move toward the hatch, saying, "This is the saddest moment of my life."

Rather than worrying about astronauts not wanting to come back in, the space agencies should have been worrying about them not being able to. It took White twenty-five minutes to get back through the hatch and safely in the spacecraft. Not helping his general state of mind was the knowledge that—should he run out of oxygen or pass out for any other reason—McDivitt was under orders to cut him loose rather than risk his own life trying to wrestle White back through the hatch.

Alexei Leonov is said to have sweated away 12 pounds in a similar struggle. His suit had pressurized to the extent that he could not bend his knees and had to go in head first, rather than feet first, as he had trained for. He got stuck trying to close the hatch behind him and had to lower his suit pressure to get back in—a potentially lethal move, akin to a diver ascending too quickly.

The NASA History Office account includes an intriguing Cold War detail: Leonov, it claims, had been given a suicide pill in case he couldn't get back in and crewmate Pavel Belyayev was forced to "leave him in orbit." Given that death from cyanide, the poison most commonly associated with suicide pills, is slower and more ghastly than death from having one's oxygen supply cut off, there would have been little call for the pill. (As brain cells die from oxygen starvation, euphoria sets in, and one last, grand erection.)

Space physiology expert Jon Clark told me the suicide pill story is most likely untrue. I had emailed Clark at his office at the National Space Biomedical Research Institute regarding the perplexing logistics of pill-popping in a spacesuit,* and he did some asking around. His Russian sources also dismissed another rumor, that Belyayev was under orders to shoot Leonov if he couldn't get back in. In fact, it was Leonov and Belyayev's wayward landing, inside the territory of a pack of lurking wolves, that resulted in the addition, at least for a while, of a lightweight pistol to the cosmonauts' wilderness survival gear.

After Ed White's spacewalk, reports of space euphoria were

---

* It would have had to be affixed to a holder inside the helmet, just as in-helmet snack bars are. The snack bar, made of the same stuff as Fruit Roll-Ups, is positioned so that astronauts can simply bend their head down and take a bite. Or, as astronaut Chris Hadfield told me, bend their head and smear it on their face. The fruit bars are mounted alongside the drink tube, which tends to leak a bit, turning the fruit into a "gooey mass." "We just stopped using them," Hadfield said.

rare, and soon the psychologists stopped worrying. They had something new to worry on: "EVA height vertigo." (EVA is short for "extravehicular activity," meaning spacewalking.) The image of Earth rushing by some 200 miles below can cause paralyzing fear. Mir astronaut Jerry Linenger wrote in his memoir about the "dreadful and persistent" feeling that he was "plummeting earthward . . . at ten times or a hundred times faster" than he'd experienced during parachute free falls. Which he was. (The difference, of course, is that the astronaut is falling in a huge circle around Earth and doesn't hit the ground.)

"White-knuckled, I gripped the handrail . . . ," wrote Linenger of his agonized moments on the end of Mir's 50-foot telescoping arm, "forcing myself to keep my eyes open and not scream." I once listened to a Hamilton Sundstrand suit engineer tell the story of an unnamed spacewalker exiting the hatch and then turning to wrap both spacesuited arms around a colleague's legs.

Charles Oman, a space-motion-sickness and vertigo expert at the National Space Biomedical Research Institute, points out that EVA height vertigo is not a phobia, but a normal response to the novel and terrifying cognitive reality of falling through space at 17,500 miles per hour. Be that as it may, astronauts are disinclined to share. "There's a big reporting problem," says Oman.

Astronauts train for spacewalks by putting on their EVA suits and rehearsing their moves while floating in a giant indoor pool called a neutral buoyancy tank. Floating in water is not exactly like floating in space, but it's a decent simulation for the purposes of practicing tasks and gaining familiarity with the outside of your spacecraft. (Mock-ups of parts of the ISS exterior lie submerged like a shipwreck on the floor of the pool in Houston.) But the training does nothing to prevent EVA height vertigo. Virtual reality training may help to a degree, but in the end, you can't effectively "sim" the sensation of free fall in space. To get a very mild sense of what it's like, climb a telephone pole (while wearing a safety harness), and then try to stand up on the flat, pie-sized top of the pole—as self-empowerment seminar attendees and phone company applicants are sometimes made to do. "Phone companies lose about a third of their trainees in the first few weeks," says Oman.

THESE DAYS, THE PSYCHOLOGISTS have turned their attention to Mars. The breakaway effect appears to have been repackaged as "earth-out-of-view phenomenon":

> In the history of human beings, no one has ever been in a situation when Mother Earth, and all of her associated nurturing and comforting aspects . . . has been reduced to insignificance in the sky. . . . It seems possible that it will induce some state of internal uncoupling from the Earth. Such a state might be associated with a broad range of individual maladaptive responses, including anxiety and depressive reactions, suicidal intention, or even psychotic symptoms such as hallucinations or delusions. In addition, a partial or complete loss of commitment to the usual (Earth-bound) system of values and behavioral norms may occur.

The passage is from the book *Space Psychology and Psychiatry.* I read it aloud to cosmonaut Sergei Krikalyov. Krikalyov is a veteran of six missions and now the head of training at the Yuri Gagarin Cosmonaut Training Center in Star City, the town outside Moscow where cosmonauts and other Russian space professionals and their families live and work.

Krikalyov isn't a man who snorts, but his response implied one: "Psychologists need to write papers." He told me that in the early days of the railway system, there was concern that people would be driven insane by the sight of trees and fields rushing past through the windows. "There was suggestion to build fences on both sides of railroad, otherwise the passengers are going to be crazy. And no one was talking about this except psychologists."

Every now and then, you do come across astronauts who describe an anxiety unique to space. It's not fear (though apparently astrophobia,* fear of space and stars, does exist). It's more of an intellectual freak-out, a cognitive overload. "The thought of one hundred trillion galaxies is so overwhelming," wrote astronaut Jerry Linenger, "that I try not to think about it before going to bed, because I become so excited or agitated or something that I cannot sleep with such an enormous size in my mind." He sounded a little agitated just writing that sentence.

Cosmonaut Vitaly Zholobov described looking at a star while on board the Soviet space station Salyut 5 and grasping in a sudden and visceral way that space is a "bottomless abyss," and that it would take thousands of years to travel to that star. "And that's not the end of our world. One can travel further and further and there is no limit to that journey. I was so shocked that I felt something crawling up my spine." The 1976 mission was terminated early due to what one space history journal article termed "psychological/interpersonal difficulties."

Zholobov lives in the Ukraine, but my indomitable Russian interpreter Lena tracked down his crewmate, Boris Volynov. Volynov is seventy-five now and lives in Star City. Lena telephoned him to see if he might have time to chat. The call was short. There were psychological/interpersonal difficulties.

"Why should I talk to her?" said Volynov. "So she can sell a lot

---

* One self-help phobia Web site helpfully reassures the afflicted that "if you have no plans to travel into space . . . astrophobia may not significantly impact your life."

of books and make a lot of money off me? She will use me like a milking cow."

"Then I'm sorry to have bothered you, Boris," said Lena.

Volynov paused. "Call me when you get here."

THE COSMONAUT HAS GONE grocery-shopping. Lena and I are meeting Volynov at the restaurant upstairs from the Star City market, where he is picking up some items for a visit with his grandchildren. From our table on the restaurant veranda, we can see the high-rise apartment buildings and training facilities. At a mile and a half square, Star City is more a town than a city. ("Starry Township" is a closer though less snappy translation.) It has a hospital, schools, a bank, but no roads. The buildings are linked by cracked bitumen sidewalks and dirt paths through fields of wildflowers and pine and birch forest. The passport control station smells like soup. There's fabulous Soviet-era sculpture in the lobbies and courtyards and space-themed mosaics and murals on the walls. I find it charming. U.S. astronauts who train here before riding the Soyuz capsule back from the International Space Station often do not. With charm comes charm's sidekick, dilapidation. Stairs are worn and chipped. Patches of stucco have dropped off the façade of the grocery store, as though it had been shelled. When I went off to use the ladies room at the museum, an employee ran after me, waving a pink bloom of scrunched toilet paper, for there were no dispensers.

I spot Volynov through the uprights of the patio railing. He has broad Soviet shoulders and a spectacular, undiminished head of hair. He doesn't move like most seventy-five-year-old men. He *strides*, leaning forward with purpose and solidity (and groceries). He is wearing his medals. (Cosmonauts are awarded the Hero of the Soviet Union star upon completing a mission.) I will learn later than Volynov was bumped from his first mission assignment when the state discovered that his mother was Jewish. Though he trained with Yuri Gagarin, he would not fly until 1969.

Volynov orders tea with lemon. Lena tells him I am interested in hearing about Salyut 5—what happened? Why did he and Zholobov come down early?

"On the forty-second day," Volynov begins, "it was an accident. Electricity switched off. No light, everything stopped, all engines, all pumps. Dark side of orbit. No light from the windows. Weightlessness. We don't know where is floor or ceiling or maybe it's a wall. No new oxygen coming. So you may count only on the volume in the station. Nobody from Earth could hear us, and we didn't have any connection with them. Many problems. Hair like this." With both hands, Lena mimes pulling her hair out. "What to do? Finally we start to fly over transmitters and we could talk to ground. They told us . . ." Volynov laughs at the memory. "They told us to open the instruction book to page number such and such. Of course this is of no help. We restored the station with our heads and hands. It took us one hour and half.

"After that, Vitaly couldn't sleep anymore. He started to have headache, terrible headaches. Stress. We've eaten all the medicine. On the ground, they were worried about him. They ordered us to come down." Volynov says he worked on his own for thirty-six hours without sleep, readying the descent module. It would seem that Zholobov suffered some sort of breakdown.

Later in the afternoon, Lena and I go for a walk through the pines with the psychologist to the cosmonauts, Rostislov Bogdashevsky. He has been at Star City for forty-seven years. Much of what he tells me is abstract and opaque. My notes say things like, "self-organization of dynamic structures of interpersonal relations in human society." But what he had to say about Volynov and Zholobov was clear and simple. "They were exhausted by overwork. The human organism is built for tension and relaxation, work and sleep. The principle of life is rhythm. Who out of us can work nonstop for seventy-two hours? They made them sick people."

Neither Volynov nor Bogdashevsky spoke of interpersonal difficulties on board Salyut 5. If anything, the mission seems to have brought the men closer, in the way that disaster and near-death will do. Volynov recalls the rescue helicopter's approach. "Vitaly heard it first. He tells me: 'Boris, there are people who are your relatives due to blood connection. But there are also people who are your relatives due to things you do together. Now you are closer to me than your brother or sister. We landed. We are alive. The prize is life.' "

When Volynov hears that Lena and I have been to the Star City museum, he tells us that on a later mission, he returned to Earth in a Soyuz capsule identical to the one on display. "I can still fit," he says. I try to picture it—Volynov in his business suit, squeezing himself inside the placental confines of a Soyuz seat.

His own capsule, Soyuz 5, isn't displayed because it was badly damaged. It did not separate properly from the rest of the Soyuz spacecraft, began to tumble, and reentered the atmosphere backward. Volynov, traveling alone, bounced around "like a ping-pong ball." Only one side of the capsule has a heat-resistant coating, so the outside charred and the inside began to bake. Rubber on the hatch seal was burning. "You could see big balloons because of the heat."

"Balloons?"

Lena consults Volynov, then turns back to me. "When you bake potatoes in an open fire, you see the same things on the potatoes. Foam? Bubbles."

"Blisters!"

"*Da, da, da*. Blisters."

Volynov waits for us to finish. "My spaceship looked like these potatoes." There was a noise like a train, he says. "I thought the floor was opening under my feet, and I had no pressure suit; there is not room for it. And I thought, 'That's it. It's the end.'" If the capsule hadn't eventually broken free and stabilized in the proper position, Volynov would have been killed.

"When the helicopter arrived, I ask the crew, 'Is my hair white?'"

To the first space travelers and the men responsible for their survival, mental health was low on the fret list. There was too much else to worry about.

The Hero of the Soviet Union takes a comb from his pocket. He raises his arms and sets them, like a conductor poised to cue the overture. He pulls the comb through the glorious hair (which was not then but is now white), and bends to pick up his groceries. "And now I must run away. People are waiting for me."

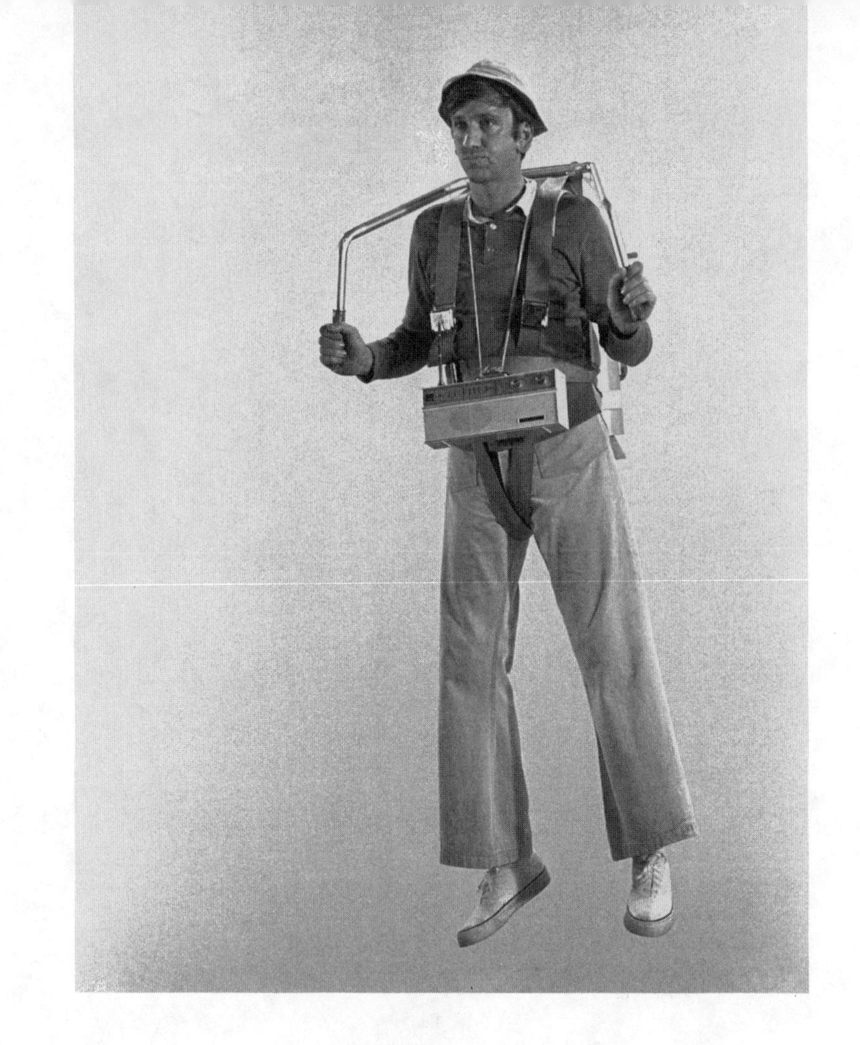

YOU GO FIRST

The Alarming Prospect of Life Without Gravity

The world's first rocket was built by Nazis to deliver bombs without leaving home. For all its fire-breathing bluster, a rocket is simply a means of delivering something—very fast and very far. The rocket was called the V-2. The first payload was the evil sleet of warheads that came down on London and other Allied cities during World War II.

The second was Albert.

Albert was a nine-pound rhesus monkey in a gauze diaper. In 1948, more than a decade before the world had heard of Yuri Gagarin or John Glenn or Ham the astrochimp, Albert became the first living creature to be launched on a rocket to space. As part of the spoils of war, the United States had taken possession of three hundred train carloads of V-2 rocket parts. They were by and large the playthings of generals, but the V-2s caught the imagination of a handful of scientists and dreamers, men more interested in the going-up than the coming-down.

One of them was David Simons. In his oral history, Simons

describes a conversation with his boss, James Henry, at the Aeromedical Research Laboratory at Holloman Air Force Base, near White Sands Proving Ground in New Mexico. The conversation is classic 1940s, an era when people regularly began their sentences with "Why, . . ." and "Boy, . . ."

Dr. Henry starts it off. "Dave, do you think man will ever go to the moon?" I like to picture him in a lab coat, pensively poking his chin with the eraser end of a No. 2 pencil.

Simons replies without hesitating. "Why, of course. It's just a matter of engineering design and time to work out the problems—"

Henry cuts him off. "Well, what would you think of having an opportunity to help us put a monkey in a captured V-2 rocket that would be exposed to about two minutes of weightlessness and measure the physiological responses to weightlessness?" It was a very long question.

"Oh! What a wonderful opportunity! When do we start?"

It is a moment that, to me anyway, signals the birth of American space exploration. It captures both the geeky excitement and the hand-wringing uncertainty over what might befall a human organism shot to the edges of the known world. Space was an environment in which no one and nothing on Earth had evolved, or, for all the scientists then knew, could survive.

Henry put Simons in charge of Project Albert. I'm looking at a book with photographs from the project. There is the V-2 poised for flight, 50-plus feet tall. There is Albert, with his rhesus monkey muttonchops and delicate eyelids cast down like a doll's. Below this, a shot of Albert strapped to a tiny stretcher, being slid inside a makeshift aluminum capsule that will fit into the nose cone where warheads were meant to go. You can't see the face of the enlisted man who holds him, just his midsection: the fly of his khaki pants and the cuff of a too-short shirtsleeve. His nails are dirty. There is his wedding ring. What does his wife think? What does he think? Does it strike him as odd: launching this towering rocket, the world's first ballistic missile, with nothing on board but a doped-up monkey?

Probably not. Aerospace professionals at the time were gripped with almost universal foreboding at the prospect of cutting loose from gravity's hold. What if man's organs depended on gravity to function? What if the pumping of his heart failed to push his blood through his veins, and instead merely churned it in place? What if his eyeballs changed shape and compromised his visual acuity? If he cut himself, would his blood still coagulate? They worried about pneumonia, heart failure, debilitating muscle cramps. Some fretted that without gravity, signals from floating inner-ear bones and other cues to the body's position would be absent or contradictory—and that this might cause perturbations that would, to quote aerospace medicine pioneers Otto Gauer and Heinz Haber, "deeply affect the autonomic nervous functions and ultimately produce a very severe sensation of succumbence associated with an absolute incapacity to act." I queried an online dictionary about *succumbence*. It said, "Did you mean *succulents*?"

The only way to know was to send a "simulated pilot" up there—to launch an animal in the nose of a thundering V-2 rocket. The last attempt at something similar took place in 1783. That time the experimenters were Joseph and Étienne de Montgolfier, the inventors of the hot-air balloon. It was like something from a children's book. A duck, a sheep, and a rooster went for a ride beneath a beautiful balloon, in the skies over Versailles on a summer afternoon. On they sailed, over the king's palace and the courtyard filled with waving, cheering men and women. In fact, it was an ingenious, controlled inquiry into the effects of "high" (1,500 feet) altitude on a living organism. The duck was the control. Since ducks are accustomed to such altitudes, the brothers could assume that any harm that befell one was likely to have been caused by something else. The balloon landed uneventfully after a two-mile voyage. "The animals were fine," reads Étienne de Montgolfier's report of the flight, "and the sheep had pissed in the cage."

Gravity turned out to be the least of the Alberts' concerns. There were six Alberts, told apart, like kings or movie sequels, by the Roman numeral after their name. It was Albert II who made history. (Albert I suffocated while awaiting liftoff.) The excellent volume *Animals in Space* reproduced the historic printout from the recorder that monitored Albert II's heart beats and the breaths he took during the zero-gravity portion of the flight, 83 miles high. They did not stray far from normal. (He had, like all the Alberts, been anesthetized.) They were also among his last. The nose cone tore loose from its parachute and fell to the desert floor. At worst, a lethal scenario. At best, a very severe sensation of succulents. The National Archives has footage of Albert II's launch and flight. I didn't order a copy. The shot list was enough.

> CU [CLOSE-UP]: . . . Several scenes of little monkey being prepared for flight in V-2, being placed in box with head sticking out, given hypodermic . . .
>
> Night shot, launching of V-2.
>
> CU: Parachute rolled up into ball on ground.
>
> CU: Smashed instruments and equipment in warhead.
>
> CU: Remains of the section containing the monkey.

AT FIRST BLUSH, Project Albert is difficult to fathom. Here are men contemplating sending a human being into space atop a tank carload of explosive chemicals, and they're worried he might be harmed by *gravity?*

To understand the Project Albert mind-set, you need to spend a few moments pondering the forces of gravitation . If you are like me, you have tended to think of gravity in terms of minor personal annoyances: broken glassware and sagging body parts. Until this week, I failed to appreciate the gravitas of gravity. Along with electromagnetism and strong and weak nuclear forces, gravity is one of the "fundamental forces" that power the universe. It was reasonable to assume that gravity might have something darker up its sleeve that mankind was yet unaware of.

A quick refresher: Gravity is the pull, measurable* and predictable, that one mass exerts on another. The more mass involved, and the shorter the distance between the masses, the stronger the pull. The moon is more than 200,000 miles away, yet it is massive enough that without any conscious effort, without plugging anything in, it pulls the Earth's water and even its tectonic plates moonward, causing ocean and (very, very small) land tides. (Earth exerts similar forces on the moon.)

Gravity is why there are suns and planets in the first place. It is practically God. In the beginning, the cosmos was nothing but empty space and vast clouds of gases. Eventually the gases cooled to the point where tiny grains coalesced. These grains would have spent eternity moving through space, ignoring each other, had gravitational attraction not brought them together. Gravitation is the lust of the cosmos. As more particles joined the orgy, these

---

* Using—how cool is this?—a gravity meter. Walk over an area of very dense rock while holding one of these meters, and you can watch the pull of gravity increase. (Fluctuations in Earth's density change its gravity enough to pull missiles off their trajectory by as much as a mile or so; gravity maps of Earth were once top-secret Cold War possessions.) This effect is lessened if the dense rock is a tall mountain and you're four or five miles above the mean surface of Earth. If you carry a bathroom scale to the top of Mt. Everest, you may see that you actually weigh a tiny bit less, not counting the marbles you have obviously lost.

celestial blobs grew in size. The bigger they became, the bigger the pull they exerted. Soon (in a thousands-of-centuries sort of way) they could lure larger and more distant particles into the tar pit of their gravitational influence. Eventually stars were born, objects big enough to pull passing planets and asteroids into orbit. Hello, solar system.

Gravity is the prime reason there's life on Earth. Yes, you need water for life, but without gravity, water wouldn't hang around. Nor would air. It is Earth's gravity that holds the gas molecules of our atmosphere—which we need not only to breathe but to be protected from solar radiation—in place around the planet. Without gravity, the molecules would fly off into space along with the water in the oceans and the cars on the roads and you and me and Larry King and the dumpster in the In-N-Out Burger parking lot.

The term "zero gravity" is misleading when applied to most rocket flights. Astronauts orbiting Earth remain well within the pull of the planet's gravitational field. Spacecraft like the International Space Station orbit at an altitude of around 250 miles, where the Earth's gravitational pull is only 10 percent weaker than it is on the planet's surface. Here's why they're floating: When you launch something into orbit, whether it's a spacecraft or a communications satellite or Timothy Leary's remains, you have launched it, via rocket thrust, so powerfully fast and high and far that when gravity's pull finally slows the object's forward progress enough that it starts to fall back down, it misses the Earth. It keeps on falling around the Earth rather than to it. As it falls, the Earth's gravity keeps up its tug, so it's both constantly falling and constantly being pulled earthward. The resulting path is a repeating loop around the planet. (It is not endlessly repeating, though. In low Earth orbit, where spacecraft roam, there's still a trace of atmosphere, enough air molecules to create a teeny amount of drag and—after a couple years—slow a spacecraft* down enough that without a rocket engine blast it falls out of orbit.) In order to escape the Earth's gravitational pull completely, an object must be hurtling at Earth's escape velocity: 25,000 miles per hour. The more massive a celestial entity, the harder it is to break its hold. To escape the monstrous gravity of a black hole (a huge collapsed star), you'd need to travel faster than the speed of light (about 670 million miles per hour). In other words, even light can't escape a black hole. That's why it's black.

Getting back to weightlessness. Weight is a bit of a mind-bender. I had always thought of my weight, on any given day, as a constant, a physical trait like my height or my eye color. It's not. I weigh 127 pounds on Earth, but on the much smaller moon, whose gravitational pull is one sixth of Earth's, I weigh about as much as a beagle. Neither weight is my real weight. There is no such thing as a real weight, only real mass. Weight is determined by gravity. It's a measure of how fast you'll accelerate if you happen to be dropping through the air like Newton's apple. (Here on Earth, were there no atmospheric drag to slow you down, gravity would accelerate you at the rate of 22 miles per hour faster for each second that you fall.) If you're standing on the ground, you obviously don't speed up, but the pull is still there. You're not falling, just pressing. The acceleration reads as weight on a bathroom

* Or a space station garbage bag or a NASA spatula. When astronauts let go of objects, they become satellites for the few weeks or months it takes them to lose speed and fall out of orbit. The term "satellite" applies to any object orbiting the earth. The "spat sat," as the orbiting spatula was known, had been used to test a spackling technique to fix dings in the exterior of the Space Shuttle caused by, ironically, orbiting debris. You don't have to worry about being killed by falling spatulas or LSD gurus, because these things burn up when they reenter the Earth's atmosphere. (Dr. Leary was recremated sometime in 2003.)

scale. When there's nothing to press against, as in the free fall of orbit, then you are weightless. The "zero gravity" that astronauts experience aboard an orbiting spacecraft is simply a continuous state of falling around the Earth.

If something provides a supplemental source of acceleration—something added to the acceleration prompted by Earth's gravity—now your weight will change. Take your bathroom scale into an elevator and watch the readout as you take off. You will briefly gain weight, and perhaps a minor reputation around the building. The elevator's acceleration has added an extra earthward pull to the earthward pull of gravity. Contrariwise, when the elevator approaches the top floor and slows down, the deceleration renders you briefly lighter; it has accelerated you skyward, counteracting some of the Earth's downward pull.

Why is there this force, this pull between objects? Poking around on the Web for a suitably patient entity to ask, I came upon the Gravity Research Foundation, founded by multimillionaire businessman and fire alarm magnate Roger Babson. After gravity pulled Babson's sister toward the bottom of a river and she drowned, he became history's most voluble antigravity activist, publishing screeds like *Gravity: Our Enemy No. 1.* If I were Babson, I might have nominated water or currents for the number-one spot, but the man was unshakable in his ire.*

---

* To inspire future generations to take up the fight against gravity, Babson paid for stone monuments to be erected at thirteen prominent American colleges. Colby College's "antigravity stone," as it became known, states its goal as follows: "To remind students of the blessings forthcoming when a semi-insulator is discovered in order to harness gravity as a free power and reduce airplane accidents." The students were differently inspired: In what became a joyous progravity rite, the antigravity stone was knocked over so many times

Babson is dead, but the foundation lives on. It no longer characterizes its efforts as antigravity, a term that has come to connote "crackpot." "We are neither 'pro-gravity' nor 'anti-gravity,'" director George Rideout, Jr., told a journalist who profiled the organization in 2001. They are, he said, just trying to learn as much as possible about it. I contacted Rideout seeking an explanation of why gravity exists. He told me to go ask a physicist.

I did. I made a hobby of it. But *why* are two masses drawn together, I'd say. "Mary, Mary, Mary," was the kind of response I tended to get. "Because space-time exists," said one physicist. "What does 'why' mean?" said another. Perhaps gravity is a mystery even to those who understand it. I can well imagine that the prospect of messing with it must have been daunting to the pioneers of aerospace medicine out in the desert in 1948.

DISMAYED BUT UNDETERRED, Simons and his crew launched four more Alberts. Albert III's rocket exploded. Alberts IV and V were, like Albert II, victims of malfunctioning parachute systems. Albert VI made it to the ground with his vital signs little changed, but died of heat prostration while rescuers searched for the nose cone. Eventually the Air Force—and you do wonder what took them so long—stopped naming their ill-fated gravity monkeys Albert. More importantly, they began to move away

---

that the college eventually relocated it to a less prominent spot. Along with the stones, Babson left the colleges small grants but did not explicitly state that the money must go toward antigravity research. Loath to sponsor "Mickey Mouse" science, Colby used the money to erect a skyway connecting two science buildings. "At least," noted a college spokesperson, "it's off the ground."

from the V-2s in favor of a smaller, less problematic* rocket called the Aerobee.

Patricia and Michael, in 1952, were the first monkeys to survive a trip to Weightlessville. The macaques' heart rate and breathing was monitored throughout the flight and appeared to be normal. Biomedical research from this era appears to have been fixated on pulse and respiration. Publicity images from that era invariably show a physician in a white coat and crewcut, holding a stethoscope to a monkey's narrow chest. That's all the Albert papers reported on. You couldn't diagnose much from it—*yep, still alive*—but this was the limit, circa 1950, of the data you could transmit back from a rocket 30 or 50 or 80 miles up. To rule out any subtler effects of weightlessness, the Air Force would need a subject they could interview: a human. For that, they needed a safer way to go about it.

It was a team of brothers, Luftwaffe aerospace medicine pioneers Fritz and Heinz Haber, who, in 1950, dreamed up a technique known today as parabolic flight. The Habers theorized that if a pilot flies the same kind of parabolic arc as a suborbital rocket (or a baseball pop fly), then the passengers, for anywhere from 20 to 35 seconds at the top and the downward segments of the arc, will experience weightlessness, just as the monkeys had. If the pilot then pulls

---

* The V-2's directional system was notoriously erratic. In May 1947, a V-2 launched from White Sands Proving Ground headed south instead of north, missing downtown Juarez, Mexico, by 3 miles. The Mexican government's response to the American bombing was admirably laid back. General Enrique Diaz Gonzales and Consul General Raul Michel met with United States officials, who issued apologies and an invitation to come to "the next rocket shoot" at White Sands. The Mexican citizenry was similarly nonchalant. "Bomb Blast Fails to Halt Spring Fiesta," said the *El Paso Times* headline, noting that "many thought the explosion was a cannon fired for the opening of the fiesta."

out of the downward dive and heads back up and repeats the process, over and over until his fuel runs low, science will have an accumulation of several minutes of weightlessness to work with—at a fraction of the cost of building and launching rockets. These roller-coaster zero-gravity flights are still flown today by space agencies to test equipment or train astronauts or humor authors who have pestered them ceaselessly for months (more on this shortly).

Here the scene shifts to South America. The Habers had a colleague named Harald von Beckh, who lived in Buenos Aires after the war. Von Beckh knew from the V-2 and Aerobee rocket flights that weightlessness posed no grave threat to survival, but he wondered whether it would disorient a pilot or otherwise compromise his ability to fly a craft. So naturally, von Beckh went out and got some snake-necked turtles. *Hydromedusa tectifera* are, like post-war Nazis, native to Argentina, Paraguay, and Brazil. These are turtles that hunt like snakes, coiling their overlong necks into an *S* and then unwinding in bullet-fast strikes that rarely miss. That is what von Beckh planned to test. Would weightlessness put them off their game? It did. The turtles moved "slowly and insecurely" and did not attack a piece of bait placed directly in front of them. Then again, the water in which they swam was repeatedly floating up out of the jar and forming an "ovoid cupola." Who could eat?

Von Beckh quickly moved on from turtles to Argentinean pilots. Under the section heading "Experiments with Human Subjects"—a heading that, were I a doctor previously employed by Nazi Germany, I might have rephrased—von Beckh reports on the efforts of the pilots to mark *X*'s inside small boxes during regular and weightless flight. During weightlessness, many of the letters strayed from the boxes, indicating that pilots might experience difficulties maneuvering their planes and doing crossword puzzles during air battles.

The following year, von Beckh was recruited by the Aeromedical Research Laboratory at Holloman Air Force Base—home of Dave Simons and Project Albert. Simons was keen to continue his zero-gravity research using the newfangled parabolic flight technique. All he needed was a willing pilot. Only one man volunteered. Joe Kittinger "made a career" out of volunteering. "You can't get any real fun things unless you volunteer," says Kittinger in an oral history on file at the New Mexico Museum of Space History. (Kittinger has a unique sense of fun. In 1960, he volunteered to make a parachute jump into the near-airless void 19 miles above the Earth, to test survival equipment for extremely high-altitude bailouts. More on this in chapter 13.)

Kittinger would take the plane up at a 45-degree angle, and then arc it over and plunge back down, all the while watching a golf ball suspended on a string from the cockpit ceiling. "That was our instrumentation!" Kittinger told me. When the plane achieved zero gravity, the golf ball started floating. So did Kittinger, of course, but he was strapped in his seat. Meanwhile, back behind the cockpit, a Salvador Dali photo had come to life. Von Beckh and Simons were studying, among other things, cats' ability to right themselves in zero gravity. "The guys would take them and just let them float," recalled Kittinger. "Here would come a cat and I would push the cat back. A couple of times we had a monkey come floating up to the cockpit. And I would take the monkey and I would push it back."

When it became clear that a few seconds of weightlessness was more entertaining than it was troublesome, the aerospace medicine crowd began to apply their boundless nervous energy to the scenario of longer-duration missions. Would an astronaut on a three- or four-day orbit of Earth or a trip to the moon be able to eat, or did he need gravity to help the food along? How would he drink water? Does a straw work in zero gravity? Late in 1958, three captains at the U.S. Air Force School of Aviation Medicine at Randolph Air Force Base in Texas commandeered an F-94C fighter plane and fifteen volunteers and undertook a project to answer these simple questions. Though they were phrased less simply for the journal paper, which came out under the title "Physiologic Response to Subgravity: Mechanics of Nourishment and Deglutition of Solids and Liquids."

The captains were not reassured by what they found. New and never-before-encountered dangers presented themselves. Water in a cup became "an amoeboid mass" that would levitate from the cup and "envelop" the face. "The fluid flowed into the . . . sinuses as the subjects attempted to breathe. Choking—virtually a sense of drowning—was a common occurrence." Eating was deemed equally perilous. "A number of subjects reported that pieces of food hung suspended in the oropharynx and several reported that bits of food floated up over the soft palate into the nasal passages." Chewed food, they claimed, was drifting up the esophagus into the mouth, where it "caused the subjects to vomit and feel ill." I would have assumed that the vomiting was due to the plane's insane trajectory, or perhaps something having to do with zero gravity's effect on the vestibular system, but the researchers stuck to their loopy guns and coined a new, utterly nonexistent phenomenon: Weightless Flight Regurgitation Phenomenon.

Fast-forward five months. The three captains are now majors. They commandeer yet another F-94C and begin "Physiologic Response to Subgravity: Initiation of Micturition." The concern was legitimate. If you counteract the pull of gravity, will the bladder still empty correctly? Based on their zero-gravity experiences with glasses of water ("exceedingly messy"), the researchers knew better than to have the men urinate into an open container. Using scrap hosing from oxygen masks and small weather balloons, they fashioned enclosed urine receptacles. To make sure everyone needed to go, the subjects were, with characteristic Air Force zeal, told to drink eight glasses of water over the course of the two hours leading up to flight time. Severe discomfort resulted, such that several of the men had to visit the head well before the plane took off. In the end, everything worked fine, and the urine flowed normally.

Kittinger has a name for the researchers: weenies. "There were scientific papers put out all over the place by the experts that said that [zero gravity] was going to be the limit to putting man into space," says Kittinger in his oral history. "And I just sat there and laughed my butt off, because I loved it! I thoroughly enjoyed it."

You can't really blame the weenies. You have to put their concerns in the context of the times. Space and zero gravity were uncharted territory where none of the familiar rules could be assumed to apply. Over the course of history, the same sort of anxiety has appeared every time a newer, faster form of transport has come along. "When technical perfection of the steam engine made the development of railways possible, scientists were afraid that the velocity of the trains would exert harmful effects upon the human body." The quote comes from an aviation medicine text published in 1943. (Locomotives at that time could not exceed fifteen miles per hour.) In the early 1950s, as commercial flights became available, doctors feared that flying might harm the heart and adversely affect the circulation. When a Dr. John Marbarger showed that it did not, United Airlines gratefully awarded him its Arnold D. Tuttle Award.

Parabolic flights are still being flown by space agencies, but these days it's not human beings they're testing—it's equipment. Every time NASA develops a new piece of hardware—be it a pump or a heating element or a toilet—someone has to haul it up on a plane out of Ellington Field near Houston to see what sort of problems might develop in zero gravity. Twice a year, something even more problematic gets hauled up there: college students and journalists.

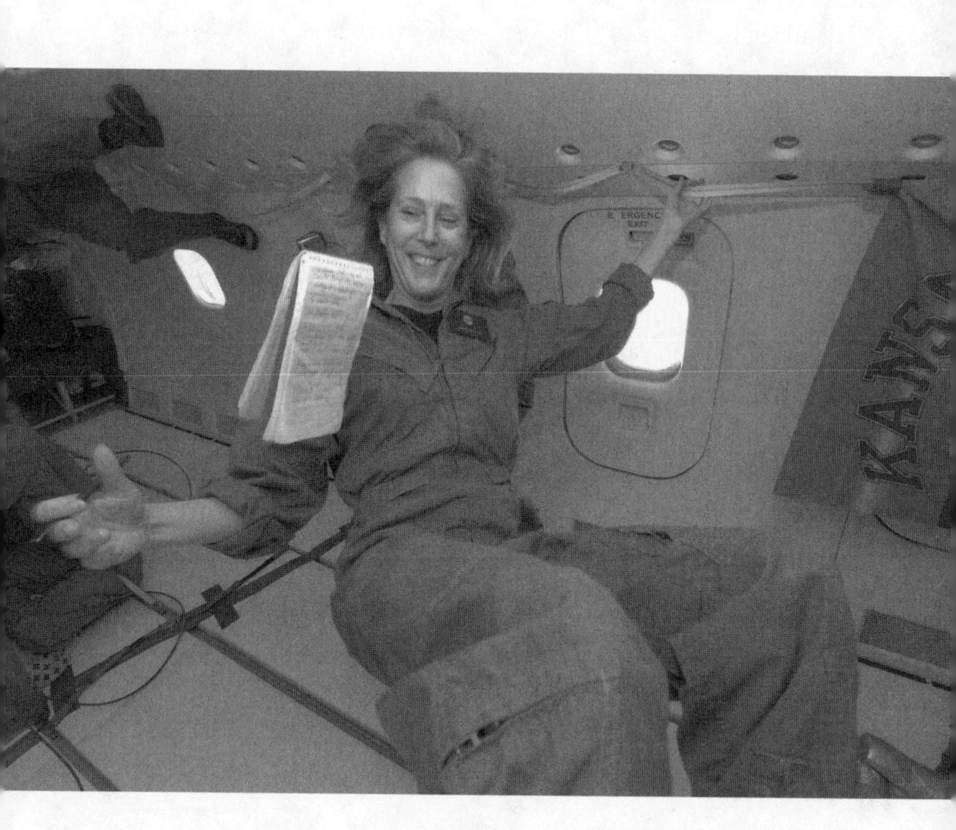

5 UNSTOWED

Escaping Gravity on Board NASA's C-9

If you stumbled onto Building 993 at Ellington Field airport, you would have to stop and wonder about the things inside. The sign on the front is as evocative and preposterous as the engraved brass one that says Ministry of Silly Walks in the Monty Python sketch of the same name. This sign says REDUCED GRAVITY OFFICE. I know what is in there, but even so, I have to stand for a moment and indulge my imagination, through which coffeepots are floating and secretaries drift here and there like paper airplanes. Or better still, an organization devoted to the taking of absolutely nothing seriously.

The real Reduced Gravity Office oversees a program whereby college and high school students compete for the chance to carry out zero-gravity research projects during a parabolic flight on a

McDonnell Douglas C-9 military transport jet.* It is run by NASA with, if anything, an excess of gravity.

I have arrived late for the safety briefing. I am signed on as the journalist for a Missouri University of Science and Technology team that is studying zero- and reduced-gravity welding. ("Reduced-gravity" refers to the situation on, say, the moon, where there is one-sixth as much gravity as on Earth, or Mars, where there's one-third. It is NASA's fondest dream to one day be welding on both.)

The safety lecturer is pointing to the wing of the C-9, now parked in the middle of the hangar where we are meeting. She has long, lank brown hair and wears a maternity blouse. "There are documented instances," she is saying, "where grown men have been pulled into the engine intake from over six feet away."† I already know this because it's in the Participant Handbook. The

---

* Some months after I visited, the flights were outsourced to the Zero G Corporation, which uses a 727. Most people just call the plane the Vomit Comet. Though NASA would like them to stop. They asked us to refer to it as the Weightless Wonder. Which pretty much makes you vomit.

† I mentioned this to an Oregon Air Guardsman I met a few weeks later. He replied that this had happened to a guy he knew. "I saw pictures," he told me, leaning forward in his seat. "He was basically leaking out the back." If you do a Google search on "Human FOD" (Human Foreign Object Damage), you will find footage of a young airman being pulled into the intake of an A-6 jet, causing sparks to shoot out the other side but not the airman himself. He appears in footage shot later that day, awake and chatting, his head bandaged but otherwise okay. A flight surgeon told me that the trick to surviving is to have your flashlight or socket wrench precede you into the maw. The object will be chewed to pieces, shutting down the engine before your head arrives on the scene. One site recommends buying neck cords for eyeglasses, lest they be pulled off one's face. It goes on to say that jet intake suction can be strong enough to "pull the person's eyeballs out," but does not recommend a product for that.

handbook uses the word *ingested,* as though the plane had played an active, sinister role in the event.

Mounted on the wall behind her is a long-handled tool reminiscent of the hooks whalers used to maneuver rafts of blubber alongside the ship. A sign identifies it as a BODY RESCUE HOOK. It is for rescuing someone who is being electrocuted in such a manner that the electricity has contracted his hand muscles, making him grip the very object that is killing him. If you try to pull him away by grabbing his arm, then your hand muscles too will contract, and now you both need rescuing. The pole is nonconductive, enabling the savvy rescuer to save a life without joining the growing conga line of electrocution victims. On this same wall, a hazard sign lists the many things that can trigger accidental discharge of the building's firefighting foam. (I once saw a video of such an event. It was like Paul Bunyan drawing a bubble bath.) Unsettlingly, "welding" is on this list.

The dangers go on and on. Hearing protection must be worn on the tarmac. We are not allowed to wear flip-flops or sandals. "Horse-play" is forbidden.

In my press materials, there's a photograph of the C-9 powering through the upward climb of the parabolic arch. It is flying at an absurd angle, the way a child moves a toy plane through the air. It seems odd to be talking about the dangers of fire-retarding suds and open-toed shoes rather than the dangers of riding a jet that repeatedly pulls out of a kamikaze dive into a climb so steep that the engines shudder.

This mix of extremes—workaday paranoia and aeronautic abandon—seems to typify the world of government-funded space travel. NASA's buildings are plastered with warning signs for the most Tinkerbell dangers. SLIP, TRIP, AND FALL hazard signs are everywhere. Honestly: everywhere. Inside the stalls in the Johnson Space Center cafeteria bathroom, the toilet paper speaks to you from a dialogue bubble printed on the dispenser: "Ladies, don't drop me on the floor. There, I could become a slip, trip, and fall hazard!" Wet-umbrella bag dispensers are installed at building entrances, courtesy of the Safety Action Team, to keep the floors dry. It's as though NASA were populated by legions of hopeless pratfalling Mr. Bean types. When a corridor makes a 90-degree turn, a block-letter sign frets, BLIND CORNER: PROCEED WITH CAUTION.

Perhaps focusing on minor workplace dangers helps space agencies cope with the very major threats they deal with on every mission: explosions, crashes, fire, depressurization. Like war, space is a formidable bogeyman that takes its victims no matter how carefully you what-if the situation. You can't control the weather or gravity, but you can control the shoes your visitor wears and the amount of water that drips onto the floor from her umbrella.

To NASA's credit, a parabolic flight has never gone down. The C-9's predecessor was the KC-135, one of which is displayed on a steel mount on the lawn outside, 10 feet up and seemingly headed for the commissary. It flew 58,000 parabolas without a "mishap."* Then again, that's the sort of thing the astronauts told themselves until the day Space Shuttle Challenger exploded 48,000 feet above the Atlantic.

It's 6 P.M. The engineering students have gone to Fuddruckers without me. I pick up some takeout and settle in for an evening of NASA TV. Because I am staying at a hotel across the street from NASA—a hotel that proudly, wordily identifies itself to callers as

* That would be a NASA Type A Mishap, as it would likely entail "injury or illness resulting in a fatality." A "mishap" as you or I might define it (say, something involving a slippery floor), is not a mishap, not even a Type D Mishap. It is a Close Call. Nonetheless, there is paperwork: the JSC Form 1257 Close Call Report Form.

"Extended Stay America Johnson Space Center"—NASA TV is the first channel that comes on. I adore NASA TV. It's often just raw feed from cameras on the space station. You'll tune in to a ten-minute shot of a solar array, immobile in the silence of space, speeding over Africa, the Atlantic, the Amazon. It calms me. I hear people at NASA say they think it's boring, and there have been efforts to slick it up with graphics and hosted programs, but much of it, thankfully, is essentially undoctored.

Today the space station astronauts finished hooking up Japan's new experimental laboratory module, Kibo. After the ribbon-cutting and press conference, there's footage of them entering the module for the first time. They're like bulls let into the ring, impelled to movement by the sudden expanse of open space. I've watched a lot of NASA TV, and you rarely see this sort of abandon. You'll see a guy hunched over a circuit board, one toe hooked under a foot restraint, bobbing gently like a boat at anchor. Or you see the crew stacked in two neat rows, facing the camera and fielding press questions. If it weren't for the floating microphone cord or someone's gold necklace levitating in front of her chin, you could easily forget they're weightless.

My noodles have gone cold because I can't look away from the TV. One astronaut is spinning horizontally, as though NASA TV had hired one of those guys who do special effects for martial arts movies. Karen Nyberg is banking like a cue ball: wall, ceiling, wall, floor. No one wears shoes, because no one's soles need to be on the floor, and even if they did, the dust and dirt don't settle there. The astronaut from Japan, Akihiko Hoshide, is crouched at the opening to the module, waiting for a clear path across the length of it. He pushes off and flies across empty air, arms in front like a superhero. I have done this in dreams. I'm in an enormous old building with 50-foot ceilings and elaborate moldings. I push off from the molding and glide across the room, then bank off the opposing wall and do it again. Whatever the dangers of parabolic flights may be, they do not dampen the anticipated joy of escaping gravity. I go to sleep feeling like a six-year-old on Christmas Eve.

When I arrive in the morning, my team's welding experiment has been loaded onto the C-9. From the outside, the plane looks like any large jetliner, but inside it has been gutted. Only six rows of seats remain, in the back. The welding device is an automated arm mounted in a glass-fronted box in a doored cabinet. The cabinet is affixed to a cart, like something a magician wheels around a stage. Two of the students and their supervisor are on their hands and knees, struggling to fit the legs of the cart into brackets mounted on the floor. The measurements are off by a fraction of an inch.

Team member Michelle Rader explains their project. Although much of what the astronauts have been doing on the space station the past decade amounts to zero-gravity construction work, things are typically bolted rather than welded. Sparks and molten metal make NASA nervous. A blob of superheated metal that drifts onto an astronaut's suit could melt through the layers and cause a leak. An enclosed and/or robotic welder is a possibility, but you first need to be sure that welding in zero gravity doesn't compromise the strength of the weld. That's what the Missouri students are testing today.

A loud crack causes heads to turn. One of the welding students has tried to force a leg into place and now it has broken. The Reduced Gravity Program manager, Dominic Del Rosso, stares at the scrum of students. His head is shaved. His arms are crossed. Do you recall Yul Brynner as the King of Siam? This is him, in a flight suit. Icy and annoyed. "What happened here?"

A small voice: "We um . . ."

Someone else takes over. "A weld broke."

The weld team points out that they did not weld the cart legs. These welds were done by someone at Missouri S&T's metal shop. Someone dials this man's number on a cell phone. There isn't anything the man can do for them, other than feel bad, which is probably all they want right now. Del Rosso doesn't care whose fault it is. He points to the exit. "Take it out of here."

Ruh-roh. Have I endured two days of NASA safety orientation briefings for nothing? Is it too late to switch teams? Do I need to start cozying up to Team Analyte Detection Via Protein Nanospores? Back in the hangar, I chat up one of the other Missouri students. He has a minor in explosives and the slightly bitter, misanthropic personality of someone who shouldn't. I ask him whether his team will still fly if they can't fix the leg.

He doesn't know. He's on the ground crew and does not get to fly. He gives me a forced smile. "It's okay." And then, remembering words that someone has told him to say: "It's an honor just to be here."

By midday, the welding unit is back on board, affixed directly to the floor of the plane. Space Weld Team is go for launch.

YOU NEVER THINK about the weight of your organs inside you. Your heart is a half-pound clapper hanging off the end of your aorta. Your arms burden your shoulders like buckets on a yoke. The colon uses the uterus as a beanbag chair. Even the weight of your hair imparts a sensation on your scalp. In weightlessness, all this disappears. You organs float inside your torso.* The result is

---

* They migrate up under your ribcage, reducing your waistline in a way no diet can. One NASA researcher called it the Space Beauty Treatment. Without gravity, your hair has more body. Your breasts don't sag. More of your body fluid migrates to your head and plumps your crow's feet. Because blood volume sensors are in the upper body only, your system thinks you are retaining too much fluid and dumps 10 to 15 percent of your water weight. (Then again, I have also heard it called Puffy-Face Chicken-Leg Syndrome.)

a subtle physical euphoria, an indescribable sense of being freed from something you did not realize was there.

If you go to the NASA Microgravity University Web page, you will see photo after photo of students concentrating intently on their projects and, in the background of many of these shots, a pair of grinning fools floating into each other like shirts in a dryer. That's me and Joyce. Joyce is from the education department at NASA headquarters in Washington. She helps run the student flights program but had never been on one of the flights. I should really be down on the floor with my team, taking notes on how it's going. I can't do this, however, because my notebook is floating in front of my face with the pages all fanned out, and I need to stare at it for a while longer. It hovers, not rising and not falling, in the manner of a party balloon a few days postparty. (When I get back to my room to review my notes, I find that I've written nothing of substance. I wasn't so much taking notes as testing my Fisher Space Pen. My notes say: "WOO" and "yippee.")

Last night on NASA TV an astronaut, in response to a schoolchild's question, said that being in zero gravity was like floating in water. Not exactly. In water, you sense the liquid's help—buoying you and supporting your weight. When you move, you feel it push back on you. You are floating, but a heaviness remains. Here on the C-9, for twenty-two seconds at a go, you are floating in air without effort, without help, without resistance. Gravity has given you a hall pass.

The thing weighing us down is Del Rosso. He has told us to hold on to a strap with one hand. This means that every time I'm floating, I reach the limits of my tether and swing around to the left. This causes me to enter the air space over the Kansas University team's electromagnetic docking rig. To retreat, I have to extend my leg down and push off the frame of it. "Don't kick their experiment!" barks Del Rosso. Like I meant to. *I hate your stupid electromagnetic docking thing, take that!* It's just that this floating business takes getting used to. You can ask Lee Morin. Mission Specialist Morin told me it takes about a week to feel comfortable floating. "Then it seems like the natural thing. To float like an angel. I don't know whether it's like you're, you know, back in the womb or something, but it's like the natural way. And it seems very odd to think about walking with shoes."

"Feet down!" yells a blue flight suit. This is our cue to bring our legs back underneath us, because gravity is coming back. It comes on gently—you're not dropped from the ceiling—but still, you don't want to come down on your head. Some of us lie on our backs during the double gravity portion, as we've heard we're less likely to become nauseated that way.

Gravity disappears again, and we rise up off the floor like spooks from a grave. It's like the Rapture in here every thirty seconds. Weightlessness is like heroin, or how I imagine heroin must be. You try it once, and when it's over, all you can think about is how much you want to do it again. But apparently the thrill wears off. "At first," wrote astronaut Michael Collins in a book for young adults, "just floating around is great fun, but then after a while it becomes annoying, and you want to stay in one place. . . . My hands kept floating up in front of me, and I wished I had pockets or somewhere to put them." Astronaut Andy Thomas told me how irritating it was to never be able to set something down. "Everything has to have a bit of Velcro on it. You're forever losing things. I brought one nail file with me on Mir, so I was very careful with it. About a month before the end of the mission, it popped out of my hand. I turned to grab it, and it was just gone. It went down with Mir. Once we lost a whole Sharps container. Big thing. Gone. We never saw it again."

There is some annoyance going on today. One team's computer keeps shutting down. It's one of those rugged laptops that protect themselves by shutting down when they detect a sudden spike in acceleration. On Earth, this means it has been dropped. Up here, it means the pilot is pulling out of the dive.

Nothing works as it's supposed to in zero gravity, or zero G, as it's also known. "Even something as simple as a fuse," astronaut Chris Hadfield told me, mistaking me for someone who knows how a fuse works. Now I know: Fuses have a metal strip that melts in response to a surplus of current. The molten bit drips away, leaving a gap that interrupts the power flow. Without gravity, the droplet doesn't drip, so the power continues to flow until the metal boils, by which time the equipment has fried. Zero gravity is part of the reason NASA price tags seem so extravagant. For every new piece of equipment that goes up on a mission—every pump, fan, throttle, widget—a prototype must be flown on the C-9 to be sure it works in weightlessness.

Overheating equipment is a common theme in zero G. Anything that generates heat tends to overheat, because there are no convection currents in the air. Normally, hot air rises—because it's thinner and lighter; the livelier molecules are all bouncing off each other and spreading out more than they do in cooler air. When hot air rises, cooler air flows in to fill the vacuum left behind. Without gravity, nothing is any lighter than anything else. It's all weightless. The heated air just sits where it is, getting hotter and hotter and eventually causing damage to the equipment.

Human machinery tends to overheat for the same reason. Without fans, all the heat that exercising astronauts generate would hang around their body in a tropical miasma. As would exhaled breath. Crew members who hang their sleeping sacks in poorly ventilated spots get carbon dioxide headaches.

In the case of the Space Weld Team, it is the human machinery that's most notably out of commission. It's not something you can fix with a fan.

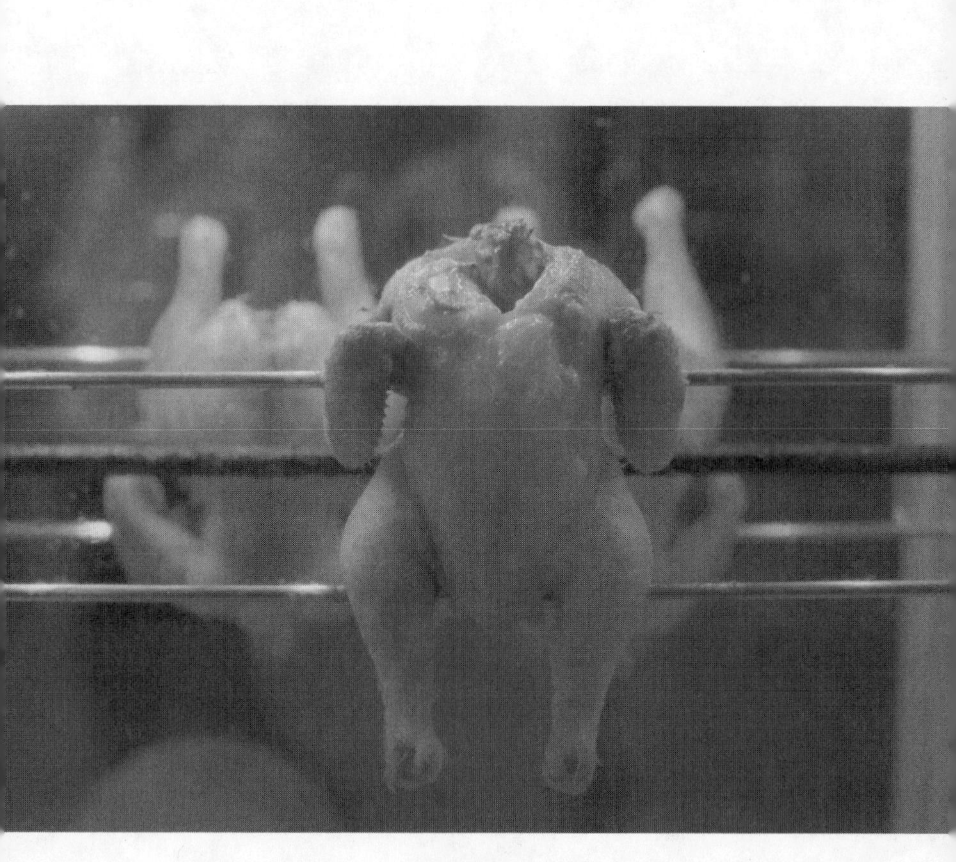

6

THROWING UP AND DOWN

The Astronaut's Secret Misery

On the ceiling of the C-9 is a red numerical display of the type you see at deli counters, telling patrons which number is being served. This one is counting parabolas, twenty-seven so far. Three more and it's over. We were told not to "go Supermanning around the cabin," but I have to break the rules. As gravity fades out on the twenty-eighth parabola, I pull up my legs, crouch on the windowpane, and then gently uncoil, launching myself across the cabin of the plane. It's like pushing off from the wall of a swimming pool, but the pool is empty and it's air you're gliding through. It's probably the coolest moment of my entire life. But not of Pat Zerkel's life. The Missouri space welder has been belted down in the front row of seats. Though weightless, he appears heavily burdened. A white bag hovers near his face. It is held open with both hands, like a hat carried through a crowd for tips.

"*OOOooulllrr-aaghchkkk, khkkk.*" Pat has been ill since the fourth parabola. At parabola number 7, the flight surgeon came over to hold him steady during the weightlessness, hoping it would help. (And to keep him, as he told me later, from "floating away helpless and vomiting everywhere.") At parabola number 12, men in blue flight suits gave Pat a shot and helped him to the back of the plane, where he would remain for the rest of the flight.

The special evil of motion sickness, the genius of its cruelty, is that, generally speaking, it hits you when you're up. A sunset sail on the San Francisco Bay, a child's first roller-coaster ride, a rookie astronaut's first trip to space.* There is no faster route from joy to misery, from *yee-ha* to *oooulllrr-aaghchkkk.*

In space, motion sickness is more than an unpleasant embarrassment. An incapacitated crew member makes for the most costly sick day in the world. An entire Soviet mission, Soyuz 10, was aborted due to motion sickness. You'd think science would have it licked by now. It's not for want of trying.

To FIGURE OUT how best to prevent motion sickness, you first need to figure out how best to bring it on. Aerospace research has excelled at the latter, if not the former, and perhaps nowhere more triumphantly than at the U.S. Naval Aerospace Medical Institute in Pensacola, Florida: the birthplace of the human disorientation device. In a 1962 NASA-funded study, twenty cadets agreed to be harnessed to a chair mounted on its side on a horizontal pole. Thus affixed, the men were rotated, rotisserie style, at up to thirty revolutions per minute. As a reference point, a chicken

---

* A journalist's ride in Tom Cruise's two-seat biplane. Cruise piloted us through a run of aerobatic stunts, the last of which, a "hammerhead," did me in. The plane had an open cockpit, and I was in the front seat, meaning that anything that might escape the "Sic Sac" that flapped in the breeze at my elbow would blow back onto Mr. Cruise's tanned and flawless face. Cruise is a cleanly man. Disaster loomed. I managed to keep my tacos down, though barely.

on a motorized spit typically turns at five revolutions per minute. Only eight of the twenty made it to the end of the experiment.

The motion sickness inducer of choice these days is the rotating chair.* Here the rider sits upright upon the seat, as if preparing to take dictation. A small motor causes the chair to spin on its base, conferring, at first glance, a joyful air to the proceedings, as though the subject had set herself awhirl—the tipsy stenographer at the office Christmas party. At the experimenter's command, the subjects, eyes closed, tilt their heads left and then right while spinning. I took a brief turn in the rotating chair that resides in the lab of space motion sickness researcher Pat Cowings, at NASA Ames. At the first head tilt, something lurched inside. "I can make a rock sick," said Cowings, and I believe her.

What has aeromedical science learned from the combined tortures of motion sickness research? For starters, we now know what causes it: sensory conflict. Your eyes and your vestibular system can't get their stories straight. Say you are a passenger belowdecks on a heaving ship. Since you are moving along with the walls and floor, your eyes report to your brain that you are sitting still in the room. But your inner ear tells a conflicting story. As the ship moves you up and down and around, your otoliths—tiny calcium pebbles that rest atop hairs that line the vestibule of the inner ear—register these movements. If the ship dips down

---

* Aerospace medicine cannot take credit for this one. Nineteenth-century insane asylums often prescribed a whirl in the Cox's chair for their more turbulent patients. Wrote one physician in an 1834 report on novel psychiatric techniques: "After having committed some irrational and spiteful act, the patient is forthwith placed on the rotating chair and revolved . . . until he becomes quiet, apologizes, and promises improvement, or until he starts to vomit." These were trying times for the mad. Alternate "treatments" included "surprise plunges into icy water."

into a trough, for instance, the otoliths rise; when the ship crests, they press down. Because the room is moving with you, your eyes detect neither. The brain gets confused and, for reasons not well understood, responds by nauseating you. Soon you are heaving too. (This is why it helps to stay up on deck, where your eyes can register the boat's motion relative to the horizon.)

Zero gravity presents a uniquely perplexing sensory conflict. On Earth, when you're upright, gravity brings your otoliths to rest on the hair cells along the bottom of the inner ear. When you lie down on your side, they come to rest on the hairs on that side. During weightlessness, the otoliths, in both situations, just float around in the middle. Now if you suddenly turn your head, they are free to ricochet back and forth off the walls. "So your inner ear says you just laid down and stood up and laid down and stood up," says Cowings. Until your brain learns to reinterpret the signals, the contradiction can be sick-making.

Given the culpability of the human otolith, it is not surprising to learn that sudden head movements are extremely, to use the lingo of motion sickness experts, "provocative." If you look at back issues of *Aerospace Medicine,* you can find pictures of grim-looking World War II troops with their heads wedged between padded vertical slats on the walls of troop transport planes: someone's attempt to stem the vomitous tide. (The smell of other people's emissions in close quarters is also highly "provocative." Cowings likes the term "inspirational.") Airsickness and seasickness were serious enough problems during the war that the government, in 1944, convened an entire United States Subcommittee on Motion Sickness. (Then again, it has also convened a U.S. Subcommittee on Poultry Nutrition and one on sedimentation.) Charles Oman, resident motion sickness expert at the National Space Biomedical Research Institute, confirmed the perils of wanton head-swiveling by mounting accelerometers on the backs of astronauts' headwear.

The ones who, just by nature, tend to jerk their heads around a lot are the ones most likely to suffer from motion sickness during a mission. What's true in space is true in a car on a winding road: No matter how much the driver behind looks like the GEICO caveman, don't whip your head around to look. According to work done by prolific 1960s motion sickness researcher Ashton Graybiel, even one head movement in highly susceptible people produces a measurable increase in their sweat level—an indication that nausea is just around the bend.*

"We actually proposed making a beeping beanie," Oman said. If astronauts moved their head too fast or too much, they'd hear a beep letting them know. Oman did not record the astronauts' responses to the beeping beanie proposal, but I'm guessing they were fairly, as they say, "provocative," for no astronautical beanie-wearing ensued. Oman did manage to get astronauts on one mission to agree to try out padded collars designed to discourage extraneous head movements, which they promptly removed. "It was perceived as an irritant," Oman said ruefully.

Astronauts have to deal with the mother of all sensory conflicts: the visual reorientation illusion. This is where up, without warning, becomes down. "You were working on a task . . . and apparently reorienting your 'down' without thinking about it, and then turning away and finding that the whole room was completely cattywampus to what you thought it was," recalls a Spacelab astronaut quoted in one of Oman's papers. (This may

---

* Intestinal activity has also been looked into as a warning bell for incipient nausea. One Space Shuttle astronaut wore a "bowel sound monitor" on his belly for the duration of the mission. Don't feel bad for him; feel bad for the Air Force security guy assigned to listen to two weeks of bowel sounds to be sure no conversations including classified information had been inadvertently recorded.

have been Pat Zerkel's problem; he told me he'd had "the distinct feeling of losing any sense of up or down.") It happens most readily in spaces with no obvious visual clues as to which is the floor and which the ceiling or wall. The Spacelab tunnel was notorious. One astronaut found traveling through it so reliably nauseating that, he told Oman, he'd sometimes pay a visit simply to make himself "get better by vomiting." Even just a glimpse of a fellow astronaut oriented differently from oneself could bring it on. "Several Spacelab crew described sudden vomiting episodes after seeing a nearby crew member floating upside down."* Nothing personal.

Experts like Oman keep changing their minds about whether drugs are a good idea. In space, as at sea, recovery is a process of adaptation; if you're under the covers in the fetal position, you're not exposing your vestibular system to the new reality. Overdoing it, on the other hand, can mean crossing the threshold and making yourself sick. Drugs help keep astronauts out of bed, moving and going about their work. But they also confer a false sense of immunity, encouraging one to overdo it. Motion sickness drugs don't make you immune; they simply raise the threshold for sickness.

For anyone taking a short trip, across the Channel or on the C-9, drugs are the answer. NASA gave us Scop-Dex (the dextroamphetamine counteracting the sedating effects of the scopolamine). Even then, most flights have at least one or two "kills," as the blue

---

* Hanging around upside down is inconsiderate to your crewmates for another reason. It's hard to understand what someone is saying when his mouth is upside down. We rely on lip-reading more than we think in everyday conversation. Astronaut Lee Morin told me that it's very hard to read someone's lips if he or she is tilted more than 45 degrees. Plus, he added, "you get the chin thing." Chins look like noses. Very distracting.

flight suits call the stricken. Pat looked queasy before the parabolas even began. It's possible he's someone who developed a conditioned response to the sight of a vehicle—in his case, a plane—that once upon a time made him horrifically ill. People who say they "get sick just looking at a boat" are not always exaggerating. (Relaxation and counterconditioning techniques can help in these cases.) People also develop conditioned responses to the smell of vomit. "This is why motion sickness can seem contagious," says Oman.

One thing the Pensacola research proved is that it helps to focus on something other than how you are feeling. The eight who finished rotisserating on the human disorientation device were those who had been given "constant mental arithmetic" tasks or timed button-pushing sequences to complete. Mental as opposed to written, because the last thing you want to be doing when you're fighting off motion sickness is reading. In particular, avoid reading papers such as "Analysis of Vomitus and Contents of Gastrointestinal Tract."

RUSTY SCHWEICKART DID everything wrong. Schweickart was an astronaut on Apollo 9, charged with testing the life-support backpack that the Apollo 11 crew would wear on their history-making stroll on the moon. Schweickart was to put it on, power it up, and head into the depressurized Lunar Module. Because he'd been sick during parabolic-training flights, he'd been exceedingly cautious the three days leading up to the spacewalk. "My whole modus operandi . . ." he said in his NASA oral history, "was to keep my head as still as possible and not to move around a lot." There's the first problem: He delayed his adaptation. On day three, Schweickart had to put on his EVA suit. This is, as he describes it, a "real contortionist challenge" with a lot of ducking down and doubling over. Problem 2: head movements. "Suddenly I had to barf, . . . and I mean, that's not a good feeling. But of course you feel better after you barf." Encouraged, he continued his preparations, moving over to the Lunar Module. Problem 3: the dreaded visual reorientation illusion. "You're used to being up, and when you go over there, it's down." When he got there, he had to wait for his crewmate to catch up to where he was on the checklist of tasks. "I've basically got nothing to do." Problem 4. "When your mind is suddenly—[its] priorities are gone, then . . . malaise gets the top priority in your brain. All of a sudden, I had to barf again."

With space motion sickness, the impulse to vomit can hit with unusual suddenness. One of Oman's Spacelab interviewees recalls sitting with a colleague who was eating an apple. "Right in the middle of it, he said, 'Aw gee!' threw the apple in the air, and vomited just like that." Launch-pad workers stuff extra vomit bags in rookies' pockets before liftoff, but even then, unfettered hurls are common.* NASA etiquette is to clean it up yourself. As one of Oman's Spacelab interviewees says, "Nobody else is going to do that work for you—and you sure don't want anybody to." Though you couldn't accuse Schweickart's fellow astronauts of a lack of compassion. Herewith, the most touching moment in the 1,200-page mission transcript from Apollo 9.

COMMAND MODULE PILOT DAVE SCOTT: Why don't you let all the rest of the powering down stuff and all that

---

* On a parabolic flight, evasive maneuvers are critical. Joe McMann, who used to run NASA's EVA Management Office, told me he was once flying with a man who threw up very abruptly. "I realized that in about three seconds, that vomit is going to come down on me in 2 G's. I was doing all kind of motions to get out of the way." One NASA employee I met swears double gravity makes it harder to throw up.

be ours, and you go get your suit off, clean up, try to eat, and go to bed?

SCHWEICKART: Okay. Cleaning up sounds pretty good.

SCOTT: Get one of those towels and wash and . . . all that stuff. That'll make you feel better.

SCHWEICKART: Okay. You want to watch the radio?

SCOTT: Yes, I'll take it.

For reasons we'll explore momentarily, NASA goes to great lengths to keep its men and women from throwing up in their helmets during a spacewalk. Schweickart and Scott had a serious talk about whether they should skip this particular EVA and just tell NASA they'd done it. Apollo 9 was a critical step in the race to put a man on the moon. The EVA life support system that Neil Armstrong and Buzz Aldrin would wear on the moon had to be tested, as well as rendezvous and docking equipment and procedures. "This is already March of 1969," recalls Schweickart in his oral history. "The end of the decade is coming right up. . . . Is this basically a wasted mission because Schweickart's barfing? . . . I mean, I had a real possibility in my mind at the time of being *the* cause of missing Kennedy's challenge of going to the moon and back by the end of the decade."

What happens if you vomit in your helmet during a spacewalk? "You die," said Schweickart. "You can't get that sticky stuff away from your mouth. . . . It just floats right there and you have no way of getting it away from your nose or your mouth so that you can breathe, and you are going to die."

Or not. U.S. space helmets, including those of the Apollo era, have air channels directing flow down over the face at 6 cubic feet per minute, so the vomit would be blown down away from the face and into the body of the suit. Disgusting, yes. Fatal, no. I ran the whole death-by-vomit scenario past Tom Chase, a senior spacesuit engineer at Hamilton Sundstrand. "There would be an extremely remote potential for any barf to get into the oxygen return duct, behind the astronaut's back," he began. "It's one of five returns, including four at the extremities, so even if one was blocked, it would be unlikely to create a complete system blockage. If it somehow did, then the crew member could shut down their fan and go on 'purge,' where they would vent out the Display and Controls Module purge valve and continue to get fresh oxygen flowing into their helmet from their pressurized tanks." Chase shut down his fan for a moment. "So you see we've really thought this one through."

Even if the vomit lingered in front of your nose and mouth, would it kill you? Unlikely. If you inhale your vomit, or for that matter anyone else's, it will trigger a protective airway reflex: you'll cough. If all goes as nature intended, the vomit will be turned away at the gates. The reason Jimi Hendrix died from inhaling his vomit (mostly red wine) was that he was so drunk that he'd passed out; his cough reflex was out of commission.

However. Vomit is a more dangerous material to inhale than, say, pond water. As little as a quarter of a mouthful can cause significant damage. The stomach acid that is a routine ingredient in vomit will handily digest the lining of the lungs. Also, vomit, unlike (hopefully) pond water, often includes chunks of recently ingested food: things to get stuck in your windpipe and suffocate you.

If stomach acid can digest a lung, imagine getting it in your eyes. "Barf bouncing off the helmet and back into the eyes would be really debilitating," says Chase. That's the more realistic danger with in-helmet regurgitation. That and vision-obstructing visor splatter.

Visor glop is a serious astronautical downer. In the words of Apollo 16 Lunar Module pilot Charlie Duke, "I tell you, it's pretty hard to see things when you've got a helmet full of orange juice." (Actually, Tang.)* Duke's in-suit drink bag began leaking† during suit checks on board the Lunar Module. (In-suit drink bags are NASA's version of the Camelbak bag.) Mission Control surmised the problem was zero-gravity-related and that it would "settle out" under lunar gravity. It did not, or not entirely. Here is Charlie Duke in the Apollo 16 mission transcript, driving on the moon, the high point of his life, as a pair of oddly named craters come into view: "I can see Wreck and Trap and orange juice."

Historically, the people who needed to worry about inhaling their vomit were not astronauts, but early surgery patients. Anesthesia, like a gallon of red wine, can both make you throw up and deaden your cough reflex. This is one reason the modern surgery patient is made to fast before the operation. In the rare event of a full stomach making its way into the operating room and disgorging its contents, doctors are equipped with an aspirator. In Hendrix's case, rescue personnel employed "an 18-inch sucker."

And you do want the model with the large-diameter suction tubing. In 1996, four physicians from the Madigan Army Medical Center in Fort Lewis, Washington, compared the time

---

* NASA didn't invent Tang, but their Gemini and Apollo astronauts made it famous. (Kraft Foods invented it, in 1957.) NASA still uses Tang, despite periodic bouts of bad publicity. In 2006, terrorists mixed Tang into a homemade liquid explosive intended for use on a transatlantic flight. In the 1970s, Tang was mixed with methadone to discourage rehabbing heroin addicts from injecting it to get high. They did anyway. Consumed intravenously, Tang causes joint pain and jaundice, though fewer cavities.

† Annoying, but probably less so than when the condom piece of his urine containment device slipped off, just before liftoff from the moon. Duke shrugged it off: "You know, warm stream down the left leg . . . and a boot full of urine."

it took to aspirate an average mouthful (90 milliliters) of simulated inhaled vomit, using first standard suction tubing and then a new, improved large-diameter model. The latter, as reported in the *American Journal of Emergency Medicine,* was ten times as fast, and less likely to suck up portions of lung.

Perhaps you are wondering what the doctors used as their "vomitus-simulating substance." They used Progresso vegetable soup. The Progresso Web site media-mention list includes *Food & Wine, Cook's Illustrated*, and *Consumer Reports*, but not, understandably, the *American Journal of Emergency Medicine.* Judging from their Web site, the Progresso people would be horrified if they knew. They have a fairly highbrow view of canned foods, even going so far as to recommend wine pairings for their product line.

Has the in-helmet upchuck ever actually come to pass? I was told that it happened to Schweickart, but my source later recanted his testimony. Charles Oman told me he knows of only one in-suit incident, and "the volume was small." It happened in the airlock of the International Space Station, while the astronaut was preparing for a spacewalk. Oman did not divulge the regurgitator's name; being sick in your spacesuit retains a stigma to this day.

Though not nearly as powerful as it was in Schweickart's day. The attitude during Apollo, Schweickart recalls, was that "motion sickness is something that weenies suffer." Cernan agrees: "To admit being sick was to admit a weakness, not only to the public and the other astros, but also to the doctors. . . ." Who might then decide to ground you. In his memoir, Cernan describes feeling sick during Gemini IX, but not letting on lest his colleagues think of him as "some nugget on a summer cruise."

Apollo 8 commander Frank Borman covered up his motion sickness. I'll let Schweickart cast the first stone: "It was well known in the astronaut corps that Frank had barfed more than once, but . . . for all kinds of reasons which are Frank's, he wouldn't really come forward with it." That left Schweickart to wear the hat that says, as he puts it, "the only American astronaut who had ever barfed in space." (Motion sickness during the Mercury and Gemini space programs was less common, probably because the capsules were extremely cramped; there wasn't enough motion for sickness.) Borman much later admitted that he was, as Cernan wrote in his memoir, "sick as a dog* all the way to the moon."

Following his flight, Schweickart dedicated himself to the study of space motion sickness. "I went over to Pensacola, and . . . I became the guinea pig, the pincushion that people stuffed their pins in and their probes in and whatnot. For six months, . . . my main job was learning as much as we could about motion sickness. And frankly, we didn't learn that much, and we don't know that much about it today, to be honest with you." The work was worthwhile in that, if nothing else, Schweickart managed to drag motion sickness out of the closet. "Rusty paid the price for us all," Cernan wrote. "Nothing was ever said in public against him, but he never flew another mission."

---

* And how sick is that? Depends on the dog, and how he's traveling. According to research done at McGill University in the 1940s, 19 percent of dogs cannot be made sick at all. In one experiment, sixteen dogs were taken out on a lake in rough weather. Two vomited in the truck on the way to the lake. Seven vomited in the boat, and one vomited both in the truck and again in the boat. Though the boat trip rendered these dogs "dejected and obviously miserable"—though perhaps no more so than the owners of the truck and boat—a later experiment with dogs on a large swing elicited much vomiting but "little subjective evidence that the dog finds the experience unpleasant." Dogs are used to study human motion sickness because the two species are about equally susceptible. Guinea pigs are not used because they, along with rabbits, are among the only mammals thought to be immune to motion sickness.

Things were said in public about Jake Garn, the astronaut-senator from Utah. Things were said in a nationally syndicated comic strip. *Doonesbury* cartoonist Garry Trudeau had been lambasting Garn's 1985 shuttle flight as a costly boondoggle. When Trudeau got wind of the fact that Garn was ill for much of the mission, one of his characters referred to "the Garn" as the unit by which space motion sickness would henceforth be measured. (In reality, there is no unit, but there is a scale, starting at "Mild Malaise" and ending at "Frank Vomiting.")

Pat Cowings laughed louder than most. When Garn was in training, Cowings had offered to teach him a biofeedback technique she developed for preventing space motion sickness. He waved her away, saying, "Yeah, I heard about that California meditation stuff. Will it grow back my hair?" (Despite what seem to me to be impressive results, Cowings to this day struggles with biofeedback's touchy-feely reputation. Even her own employer doesn't use her method. "I say to NASA: There's this big company? They're called the Navy? And they're using it now.")

No one, not Jake Garn or Rusty Schweickart or Frank Vomiting, should be embarrassed about getting sick in space. Some 50 to 75 percent of astronauts have suffered symptoms of space motion sickness. "That's why you don't see much shuttle news footage the first day or two. They're all, like, throwing up in a corner somewhere," says Mike Zolensky, NASA's curator of cosmic dust. Zolensky himself was epically sick on a parabolic flight. The only passenger worse off was the one helping astronauts practice drawing blood in zero gravity. Since his arms were strapped down, someone else had to hold the bag to his face.

Technically speaking, motion sickness is not a sickness. It's a normal response to an abnormal situation. It hits some people faster and harder than it hits others, but everyone can be made to hurl. Even fish can get seasick. One Canadian researcher recalls a story told to him by the owner of a codfish hatchery. The fishmonger had call to transport some of his tank-raised charges by sea. "After the boat had been under way for some time, all the feed they had eaten was seen to be on the bottom of the tank." The researcher listed all the species known to be susceptible to motion sickness: monkeys, chimps, seals, sheep, cats. Horses and cows can be nauseated but cannot, for anatomical reasons, throw up. There is disagreement, he said, about birds.* The author put forth that he personally had witnessed a pigeon vomiting while being spun on a rotating platform. "It is unusual," he added. I'd say.

The only humans who are predictably immune are those with nonfunctioning inner ears. It was a group of five "deaf-mutes" who failed to fall ill on a harrowing sea voyage that first alerted science to the link between motion sickness and the vestibular system. The year was 1896. Among the miserable was a physician named Minor. He states in his paper that he had heard of two other parties of deaf-mutes—twenty-two in the first group and thirty-one in the second—who regularly made long sea voyages without falling ill. Prior to Minor's paper, medical science had blamed motion sickness on lurching stomach contents and oscillating air pressure in the gut. A variety of girdles and belts were prescribed in *Lancet* articles around the time. Readers responded with their own stomach-

---

* By odd coincidence, I went to a noontime lecture today that addressed this issue. ("Turkey Vultures: Fact or Fiction?") The lecturer had brought along his pet turkey vulture, Friendly, who smelled even worse than one might imagine a turkey vulture to smell. This was, he said, because Friendly had *become sick in the car on the ride over* and vomited. Earlier he told us that turkey vultures will vomit at you if you harass them. I was in the second row, and have no trouble believing that turkey vulture vomit makes a powerful deterrent. Unless you are a coyote. Fact: The coyote considers turkey vulture vomitus a delicacy, and will harass the birds simply to get a snack.

stabilizing strategies: Singing, holding one's breath as the boat crests the swells, and "eating pickled onions freely." The rationale behind the last one being that it produces gas, which inflates the stomach and steadies abdominal pressure. The singing and flatulence perhaps explains the preponderance of deaf-mutes on ocean voyages around that time.

Ironically, NASA Ames motion sickness researcher Bill Toscano has a defective vestibular system. He didn't realize it until he rode the rotating chair. "We thought there was something wrong with the chair," says Toscano's fellow researcher Pat Cowings. I carried on a conversation with Toscano while he sat in the rotating chair, his voice rising and fading with each revolution. It's his superpower.

Since motion sickness is a natural response to a novel or sensorially perplexing motion or gravitational environment, astronauts have to go through it all over again when they return to Earth after a long mission. During the weeks or months of no gravity, their brains have been interpreting all otolith cues as acceleration in one direction or another. So when they move their head, their brain tells them they're moving. Astronaut Peggy Whitson described her first moments on Earth after coming back from 191 days on the International Space Station like this: "I stood up and the world was going around *me* at 17,500 miles per hour, as opposed to me going around the world at 17,500 miles per hour." It's called landing vertigo, or Earth sickness. (Other obscure motion sickness spin-offs include amusement park ride sickness, spectacle sickness, wide-screen movie sickness, camel sickness, flight simulator sickness, and swing sickness.)

Vile as it is, the act of vomiting deserves your respect. It's an orchestral event of the gut, complex and seamlessly coordinated: "There is a forced inspiration, the diaphragm descends, the abdominal muscles contract, the duodenum contracts, the cardia and oesophagus relax, the glottis closes, the larynx is drawn forward, the soft palate rises, and the mouth opens." Small wonder an entire "emetic brain"—or "vomiting center"—is devoted to the cause. Somewhere, I recall reading that the dinosaur formerly known as brontosaurus had a brain at the base of its tail to coordinate its lower body movements. I envisioned a brain-shaped gray organ nestled in the dinosaur's pelvis. Now I think I was wrong. Because the "emetic brain" isn't an actual brain any more than the Vomiting Center has a parking lot and a board of trustees. It's just a place in the fourth ventricle, a few clusters of nuclei a fraction of a millimeter across.

In the case of motion sickness, vomiting is an impressive lot of bother for no apparent reason. Vomiting makes sense as a bodily response to poisoned or contaminated food—gets it out of you ASAP—but as a reaction to sensory conflict? Pointless, says Oman. He says it's just an unfortunate evolutionary accident that the emetic brain happened to evolve right next to the part of the brain that oversees balance. Motion sickness is most likely a case of cross talk between the two. "Just one of God's jokes," says Pat Cowings.

IN THE 1980 London stage version of *The Elephant Man*, Joseph Merrick commits suicide* by lying down on his bed and allowing his grotesquely enlarged head to hang over the edge and crush his

* Merrick scholars disagree as to whether it was suicide or accident, but they do agree on his true first name, which was in fact Joseph, not John. The London production, as I seem to recall, used the more widely known "John," perhaps to avoid amending the program with a footnote, as I'm having to do. While I have you here, I'll tell you that David Bowie played Merrick. He wore no makeup or prosthetics and almost no clothes. He held himself crooked, just as Merrick had been, and broke your heart.

airway. It was suicide by gravity. His head had grown so heavy that his neck muscles could no longer lift it. For 20 seconds at a time, I've been feeling what that's like. When the C-9 pulls out of its downward dive to begin another climb, we are accelerated into the floor with the force of approximately 2 G's, twice the Earth's gravity. My head suddenly weighs 20 pounds, not 10. Like Merrick, I'm lying on my back—not to kill myself, but because I've been told this lowers the odds of becoming nauseated. It's very strange. I can't pick my head up off the floor of the plane.

I read somewhere that a beached whale will die from an overdose of gravity. Out of the water that normally buoys them, their lungs and body weigh so much that they collapse in on themselves. The whale's diaphragm and rib muscles aren't strong enough to expand its lungs and raise the now far heavier blubber and bone that press in on them, and the animal suffocates.

Aerospace researchers in the 1940s figured out a way to mimic excess gravity here on Earth. A rat or rabbit or chimp or, eventually, a Mercury astronaut, would be placed at the end of a long, spinning centrifuge arm. Centrifugal force accelerates body parts and fluids outwardly, away from the center of the centrifuge. As we learned and most likely forgot in chapter 4, gravity is simply your rate of acceleration. So, to mimic standing erect in excess gravity, a researcher would have subjects lie with their feet at the outside end of the spinning arm. The faster the centrifuge spins, the heavier grow the subject's organs, bones, and body fluids.

You can see what a rat's organs look like inside its body at 10 G's and 19 G's by tracking down the February 1953 issue of *Aviation Medicine* and opening to p. 54, but I don't recommend this. A team of Navy commanders at the Aviation Medical Acceleration Laboratory figured out an ingenious and horrific "quick-freeze technique," whereby anesthetized rats were immersed in liquid nitrogen while riding a centrifuge. The now nineteen times heavier blood in the heart has pooled at the bottom of the organ and weighed it down, elongating it like a wad of stretched Silly Putty. The abdominal organs are packed down into the pelvis like sandbags, the head has sunk down into the shoulders, and I don't even want to talk about the testicles. A second photograph shows the rat facing the other way around—its head at the outer end of the centrifuge arm. The extraheavy organs are now in a pileup under the rib cage, crushing the lungs and leaving the rest of the torso bizarrely empty.

The commanders were not simply entertaining themselves. The early aeromedical scientists studied the human tolerance limits for excess gravity in order to learn how to protect fighter pilots and, later, astronauts. Jet pilots are subject to as many as 8 or 10 G's as they pull out of steep dives and execute other high-speed maneuvers. Astronauts endure a few seconds of double or triple gravity during liftoff, and as many as four and sometimes more extra G's when their spacecraft reenters Earth's atmosphere on the way down. Going from the airless vacuum of space into a wall of air molecules slows their craft from 17,500 to a few hundred miles per hour. As in any abruptly slowed vehicle, the occupants are hurled forward in the direction of travel. What's dangerous about reentry is that the hurling—the period of doubled or quadrupled G forces—lasts for up to a minute, as opposed to the split-second duration of a car crash.

How many excess G's the human body can tolerate without injury depends upon how long it's exposed. For a tenth of a second, people can typically hack between 15 and 45 G's, depending on what position they're in relative to the force. When you get up into the range of a minute or more, tolerance drops alarmingly. Your heavy blood has enough time to pool in your legs and feet, depriving your brain of oxygen, and you black out. If it goes on long enough, you die. At 16 G's, wrote John Glenn of his flight-training experience on the NASA centrifuge, "it took just about every bit of strength and technique you could muster to retain consciousness." This is why astronauts lie down during reentry—so the blood doesn't pool in their legs and feet. But on your back, you are the whale on the beach. There is pain beneath the breastbone. Inhaling is a struggle. During a Soyuz reentry that went awry, ISS Expedition 16 commander Peggy Whitson endured an overly steep, overly fast reentry and a full minute in 8 G's, about double the normal hypergravity of reentry. Astronauts are taught, on the centrifuge, how to deal with this—to take quick, shallow panting breaths so the lungs never fully deflate and to inhale using the stronger muscles of the diaphragm, not the smaller muscles attached to the ribs. Even then, Whitson found it a struggle.

The human arm weighs, on average, nine pounds. That means that for the duration of reentry, Peggy Whitson's arm weighed 72 pounds. In the words of aerospace medicine pioneer Otto Gauer, "In general, only wrist and finger movements are possible above 8 G's." Meaning that an astronaut could perish because she can't raise an arm to reach a control panel. Whitson plays down the dangers. But a few weeks after I spoke to her, I met a flight surgeon who showed me photographs taken shortly after the incident. She looked, to use his word, "wasted." The next photo he showed me was of the crater in the dirt where the Soyuz capsule hit the ground. It looked like someone had tried to build a swimming pool out in the middle of the Kazakh Steppe.

Coming down is as scary as going up.

SPLASHDOWN

7

THE CADAVER IN THE SPACE CAPSULE

NASA Visits the Crash Test Lab

Crash simulation is a world comprised largely of metal and men. The simulator at Ohio's Transportation Research Center resides in a clanging, hangar-sized room with few places to sit, and none of them upholstered. The room holds little beyond the crash sled, on a track down the middle, and a few engineers in safety goggles, forever walking back and forth with coffee mugs. Other than the reds and oranges of warning lights and hazard signs, color is hard to find.

The cadaver seems almost a homey touch. Subject F wears blue Fruit of the Loom underpants and no shirt, as though he were lounging around in his own apartment. He looks deeply relaxed. As dead men do. Are. He slumps slightly in his chair and his hands rest on his thighs. Were F alive, he would not be so relaxed. In a few hours, a piston as fat as a redwood will shoot a slug of pressurized air at the seat in which he'll be strapped. Both the force of the impact and the position of the seat can be adjusted to create whatever crash scenario a researcher requires: a head-on into a wall at 65 miles per hour, say, or one car broadsiding another going 40. Today it's NASA's new Orion capsule, dropping from space onto the sea. F gets to play astronaut.

In a space capsule, every landing is something of a crash landing. Unlike a plane or the Space Shuttle, a capsule has no wings or landing gear. It doesn't fly back from space; it falls. The Orion space capsule has thrusters that can correct its course or slow it down enough to drop it from orbit, but not the kind that can be fired to soften a landing. As a capsule reenters the Earth's atmosphere, its broad bottom plows into the thickening air; the drag slows it down to the point where a series of parachutes can open without tearing. The capsule drifts down to the sea, and if all goes well, the touchdown will feel like a mild fender-bender—2 to 3 G's, 7 at most.

Touching down on water rather than earth makes for a gentler landing. The trade-off is that oceans are unpredictable. What if a cresting wave slams into the capsule as it's coming down? Now the occupants need restraints that protect them not only against the forces of being dropped straight down, but also against a sideways or upside-down landing impact.

To be sure Orion's occupants are unhurt no matter what wild card the seas present, crash test dummies and, lately, cadavers have been taking rides in an Orion seat mock-up here at the Transportation Research Center. The landing simulations are a collaboration involving the Center, NASA, and Ohio State University's Injury Biomechanics Research Laboratory.

F sits on a tall metal chair beside the piston track. Graduate student Yun-Seok Kang stands at his back, using an Allen wrench to mount a wristwatch-sized block of instrumentation on an exposed vertebra. Along with strain gauges glued to various bones on the front of the body, these instruments will measure the forces of the impact. Scans later this evening and an autopsy will reveal any injuries caused by that force. Kang was up late with yesterday's cadaver and in early this morning, but he's alert and cheerful. He has one of those happy, high-achieving personalities that self-help programs promise but rarely manage to create. He wears rectangular glasses and long bangs that march around to the sides of his head. His gloved fingers are glossy with fat. The fat—because it's slippery and because there's a fair amount of it—makes Kang's task difficult. He has been working on this mount for more than half an hour. The dead are infinitely patient.

F will be taking a hit on his lateral axis. Picture a foosball figurine—the little wooden soccer player with the skewer run sideways through his rib cage. That skewer is the body's lateral axis. Say the foosball man goes for a drive, and another car T-bones his car at an intersection. His body and organs, if he had any, would be accelerated to the left or right along that skewer. In a head-on crash or a rear-ender, they'd be accelerated along the transverse axis: from front to back, or vice versa. The third axis that researchers consider is the longitudinal—along the spine. Here the foosball player is operating a helicopter. It stalls and drops straight down to the ground. Foosball man's heart stretches down on its aorta like a bungee jumper. Should have stuck to sports.

Because astronauts are reclining on their backs during touchdown, a space capsule hitting the ocean in calm conditions creates a force on the transverse axis—front to back—by far the body's most durable. (Lying on their backs, fully supported and restrained, they can tolerate three to four times as much G force—a tenth of a second of up to 45 G's—as they could seated or standing, wherein the more vulnerable longitudinal axis takes the strain.)*

---

* Thus the best way to survive in a falling elevator is to lie down on your back. Sitting is bad but better than standing, because buttocks are nature's safety foam. Muscle and fat are compressible; they help absorb the G forces of the impact. As for jumping up in the air just before the elevator hits bottom, it only delays the inevitable. Plus, then you may be squatting when you hit. In a 1960 Civil Aeromedical Research Institute study, squatting on a drop platform caused "severe knee pain" at relatively low G forces. "Apparently the flexor muscles . . . acted as a fulcrum to pry open the knee joint," the researchers noted with interest and no apparent remorse.

Crashes often involve forces along not just one axis, but two or three of them. (Though simulations study just one at a time.) Add high seas to the capsule touchdown equation, and now you have to consider forces along multiple axes. A useful model for the kind of impact NASA must plan for—multiaxis and unpredictable—is the race-car crash. The week I visited Ohio, NASCAR's Carl Edwards, traveling at close to 200 miles per hour, slammed another car, launching his own high into the air, where it spun like a flipped quarter before slamming down into the wall. Whereupon Edwards casually got out and jogged away from the wreckage. How is this possible? To quote a recent *Stapp Car Crash Journal* paper, "a very supportive and tight-fitting cockpit seating package." Note the word choice: *package*. Safeguarding a human for a multiaxis crash is not all that different from packing a vase for shipping. Since you don't know which side the UPS guy's going to drop it on, you need to stabilize it all around. Race-car drivers are strapped tightly into custom-fitted seats with a lap belt, two shoulder belts and a crotch strap to keep them from sliding down under the lap belt. A HANS (Head and Neck Support) device keeps the head from snapping forward, and vertical bolsters along the sides of the seat keep the head and spine from whipping left or right.

Dustin Gohmert, a NASA crew survivability expert, has spent a lot of time talking to the people who design restraint systems for race cars. He and two colleagues have traveled from the Johnson Space Center to oversee the simulations this week. Gohmert has agreed to answer some questions while Kang and three other students finish instrumenting F. Gohmert has blue eyes and black hair and a lively Texas wit that he mostly sets aside while speaking into a tape recorder. He sits straight-backed and motionless while answering my questions, as though merely talking about upper torso restraints is holding him still in his chair.

Early on, NASA had dismissed race-car seats as models for Orion. For one thing, race-car drivers are sitting up, not reclining. Bad idea for astronauts who've been in space for a while. Lying down is not only safer (provided you don't have to steer); it keeps astronauts from fainting. Veins in the leg muscles normally constrict when we stand, to help keep blood from pooling in our feet. After weeks without gravity, this feature stops bothering to work. Compounding the problem is the fact that the body's blood volume sensors are in the upper half of the body. Where, without gravity, more of the body's blood tends to pool; the sensors misinterpret this as a surplus of blood, and word goes out to cut back on production. Astronauts in space make do with 10 to 15 percent less blood than they have on Earth. The combination of low blood volume and lazy veins makes astronauts lightheaded when they return to gravity after a long stay in space. It's called orthostatic hypotension, and it can be embarrassing. Astronauts have been known to faint during postmission press conferences.

There is a problem with lying on your back in a spacesuit in a very safe seat: "We threw a racing seat on its back, put a guy in it, and said, 'Can you get out?'" recalls Gohmert. "It was like putting a turtle on its back." Some months back, I watched a horizontal egress (getting out of the capsule) test of a suit prototype at Johnson Space Center. The verb "to turtle," as in "I'm kind of turtling out," was in fact used.

Getting out fast is mainly a concern when something goes wrong: The capsule is sinking, say, or it's on fire. The last time things went wrong aboard a space capsule, it was the Soyuz capsule, returning to Earth with members of the ISS Expedition 16 and 17 crews, in September 2008. (NASA has been paying the Russian Federal Space Agency to fly ISS crews home when no space shuttle is available.) The Soyuz module entered the atmosphere out of position—as it had with Boris Volynov aboard in

1969. This interfered with the aerodynamic lift that normally helps flatten its course and gentle its reentry and landing. Reentry subjected the crew to a full minute of 8 G's—rather than the customary peak of 4 G's—and a landing bump of 10 G's. The capsule landed far afield of its targeted landing site, in an empty field on the Kazakh Steppe, where sparks from the impact started a grass fire.

The Soyuz seats, like race-car seats, have side restraints along the head and the length of the torso. Which makes them safer, unless you need to get out in a hurry. "I had it all planned out," Expedition 16 commander Peggy Whitson told me in a phone interview. "I'm thinking, 'I'm going to unstrap and brace my hand here, and then lower my feet,' and of course none of that worked out. I just fell to the bottom with my head and shoulders in Soyeon's seat and my legs up and across the hatch." Gravity was not helping. "After six months, you forget how heavy things are. Like, yourself." You also, after months of weightlessness, forget how to use your legs. "Your muscles don't remember what to do." And astronauts have no pit crew to rush over and help them free of the wreckage.* Fortunately, the wind was blowing away from them and the grass fire soon burned itself out.

---

* Whitson and her crewmates, much to their surprise, did have help. Not long after touchdown, she felt someone pull her from the capsule. "I was like, 'Cool, the search-and-rescue guys are here already.' They laid me on the ground near the cesium altimeter. Which seemed odd, because we were always told to stay away from the cesium altimeter. So I start looking at the SAR guys. . . . One of them, literally, was wearing what looked like a burlap sack sewn into pants. They were Kazakh locals." One spoke some Russian. He asked Whitson's crewmate Yuri Malenchenko, "Where did this boat come from?" (The fire had consumed the parachutes.) "Yuri's like, 'No, this is a spacecraft. We were up in space.' And the guy says, '*Nu, ladna,*' which is kind of like 'Fine, whatever.'"

Worried that NASCAR-style shoulder bolsters might dangerously extend the time it takes an astronaut to get out of the capsule, Gohmert and his colleagues ran some simulations with head bolsters only. For these they used crash test dummies—or "mannequins," as Gohmert calls them, causing me to picture them taking their hits in department store outfits. It was a bad business. Gohmert described the slow-motion video footage to me. "The head stayed stationary and the body kept moving. We were actually concerned about the mannequin being okay." As a compromise scenario, the shoulder bolsters are still there but have been scaled down.

NASCAR seats are fitted to each driver, but that's too expensive to do for each astronaut. The Soyuz seats employ a compromise: a molded seat insert fit to each cosmonaut's body. But the mold still has to fit inside the seat, which ultimately limits the size of the cosmonaut. "The Russians have a much narrower range of crew sizes," Gohmert says wistfully. At the time we spoke, seats (and suits) were required to fit bodies that fall anywhere between 1st percentile female to 99th percentile male. That's 4 feet 9 to 6 feet 6, though standing height is the least of it. A seat system that supports and restrains the entire seated body has to fit buttock-knee lengths from 1st to 99th percentile, and ditto seated chest heights, foot lengths, hip breadths, and seventeen other anatomical parameters.*

---

* No one is excluded from the astronaut corps based on penis size. It is assumed that a man will fit one of the three sizes available in the condom-style urine collection device hose attachment inside the EVA suit. To avoid mishaps caused by embarrassed astronauts opting for L when they are really S, there is no S. "There is L, XL, and XXL," says Hamilton Sundstrand suit engineer Tom Chase. This was not the case during Apollo. Among the 106 items left on the moon's surface by Neil Armstrong and Buzz Aldrin are four urine collection assemblies—two large and two small. Who wore which remains a matter of conjecture.

This wasn't always the case. Apollo astronauts had to be between 5 feet 5 and 5 feet 10. It was a simple, inflexible cutoff, the governmental version of the sign by the amusement park ride: MUST BE THIS TALL TO RIDE. That meant that a lot of otherwise qualified candidates were kept out of the space program because of their stature. To today's PC-sensitized mind, that smacks of discrimination.

To Dustin Gohmert, it smacks of common sense. As things stand, NASA has to spend millions of dollars and man-hours making seats lavishly adjustable. And the more adjustable the seat, generally speaking, the weaker and heavier it is.

A further complication for the astronaut, as opposed to the race-car driver: He's got vacuum cleaner parts attached to his suit*—hoses, nozzles, couplings, switches. To be sure the hard parts of a suit don't injure the soft parts of an astronaut in a rough landing, F will be wearing a suit simulator: a set of rings duct-taped in place around his neck, shoulders, and thighs. The rings are facsimiles of the mobility bearings, or joints, of a spacesuit. (Tomorrow's cadaver, presently thawing,† will be wearing a vest with "umbilicals"—life support hoses and couplings—mounted on it.) One specific concern today is whether, on a sideways touch-

---

* And a diaper. Though the lack of a diaper doesn't mean race-car drivers don't pee in their suits. "People do it all the time," reported Danica Patrick in an interview in *Women's Health*. Except Danica. "I tried last year." She explains that this was during, appropriately enough, a yellow flag (the signal to slow down and follow the pace car, usually because of an accident). "I was like, '. . . Just do it.'" No Nike sponsorship for Danica!

† How do you tell when a cadaver is done defrosting? Bolte sticks a temperature transducer down the trachea. When the internal temperature passes 60 degrees, it's ready. Lacking that, a "thermometer up the rectum" will give you a good idea, as will moving the arms and legs to see if the joints move freely. Two to three days (in a refrigerator, please) usually does it.

down, a mobility bearing might collide with the seat's shoulder bolster and be driven into the astronaut's arm with enough force to break a bone.*

Gohmert explains how ring joints work, how they enable an astronaut to raise an arm. A pressurized spacesuit is a heavy-duty body-shaped balloon—almost more of a tiny inflated room than an article of clothing. Fully pressurized, it's all but unbendable without some sort of joints. The current suit prototype has metal shoulder rings that twist back and forth against each other, enabling astronauts to rotate their entire arm up and down, like old-fashioned doll arms. This is my analogy, not Gohmert's. Earlier in the conversation, I likened NASA's differently sized, individually selected spacesuit components to the recent development of mix-and-match bikini bottoms and tops. "I haven't bought one," Gohmert was careful to point out, "but that sounds right."

JOHN BOLTE ISN'T 99th percentile, but he's pretty big. When he drove my crappy little rental car, I swear he had to hunch forward over the steering wheel to fit in it. He was reading texts as he drove, getting updates on the score of his older son's ball game.

---

* Beware the hard things in between. The April 1995 issue of the *Journal of Trauma* includes a case report of a man whose pipe was between his BMW's airbag and his face when the bag deployed. A piece of the stem shot into his eye, resulting in "a ruptured globe." The author, a Swiss physician, has a keen globe for detail, noting that "there was tobacco all over the floor" and that the injury was similar to those seen "after a thrust of a pointed cow horn." The paper concludes with an exhortation to "behave appropriately"—no "drinking from cups, . . . holding articles on the lap, or wearing spectacles while driving." Not to thrust too pointed a cow horn, but wearing one's eyeglasses while driving surely prevents more injuries than it causes.

I was relatively certain that if he ran off the road, the car would crumple around him and he'd step from the wreckage unfazed, going "Bottom of the eighth, nine to three!"

Bolte has just arrived from OSU, where he runs the Injury Biomechanics Research Laboratory. He's here to check his students' work and to help with last-minute preparations before the piston fires. He wears hospital scrubs and a backward baseball cap. He is helping to dress F, pushing the dead man's fist through the bunched-up sleeve of a long-underwear shirt, a task he likens to dressing his five-year-old.

Now the challenge is to get F into the seat on the sled. Think of wrestling a comatose drunk into a taxicab. Two students hold F's hips, and Bolte has his hands beneath F's back. F lies on his back with his bent legs raised, like a man whose dinner chair has tipped over.

The piston is off to F's right; he'll be impacted along his lateral axis. "Lateral crashes are very deadly because . . ." Gohmert stops. "I shouldn't say crash." "Landing pulse" is the preferred NASA phrasing. (NASCAR is partial to "contact.") "NASA must train these guys," Bolte marveled at one point. "You ask them a question and you see them pause and think through their answer." Bolte isn't like that. My favorite line of the day so far has been Bolte's: "Is he leaking badly from anything major?"

What's so deadly about lateral "pulses"? Diffuse axonal injury. When an unsecured head whips from side to side, the brain gets slammed back and forth against the sides of the skull. The brain is a smushable thing. It alternately compresses and stretches out as this happens. In a lateral impact, as opposed to a head-on, the stretching pulls on the long neuron extensions, called axons, that connect the brain's circuits across the two lobes. The axons swell, and if they swell too much, you may go into a coma and die.

A similar thing happens to the heart. A heart, when it's full of blood, can weigh a good three-quarters of a pound. In a side impact, as opposed to a head-on, there's more room for it to whip back and forth on the aorta.* If the aorta stretches far enough and the heart is heavy with blood at that moment, the two may part ways. "Aortal severation," as Gohmert put it. This happens less often in a head-on collision, because the chest is relatively flat in that direction; the heart is more sandwiched in place. Hearts also come off their stalk in longitudinal impacts, like those that happen in helicopter drops, because there's lots of room for them to pull downward and exceed the limits of the aorta's stretch.

F is finally ready. We've moved upstairs to watch the action from the control room. A bank of overhead lights comes on with a dramatic *phumph.* The actual impact itself is anticlimactic. Because it is air† that's doing the impacting, sled tests are unexpectedly quiet, crashes without a crash. And they are fast, too fast for the eye to register much of anything. The video is shot at ultrafast speed, so that it can be played back in extremely slow motion.

We all lean in to see the screen. F's arm bends up underneath the shoulder bolster, the space where the rib bolster had been removed. The arm appears to have an auxiliary joint, bending where arms shouldn't bend. "That can't be good," says someone. This has been a recurring problem. As Gohmert puts it: "Gaps in

---

* How much does it move? Enough that you can sometimes feel it. In one Apollo-era study of sudden deceleration (stopping fast), five out of twenty-four subjects complained of what the researcher called "abdominal visceral displacement sensation."

† Does this sound gentle? It is not. Recall Javier Bardem in *No Country for Old Men.* If you missed the film, think of pork workers described in a *MedPage Today* article as using jolts of compressed air to force pig brains out of heads. "This 'emulsifies' the brain tissue," explained a source.

the seat tend to get filled in by body parts." (The arm will turn out not to be broken.)

F endured a peak impact of 12 to 15 G's—right on the cusp of injury. Gohmert explains that the extent of an accident victim's injuries will depend not only on how many G's of force there were, but on how long it takes the vehicle to come to rest. If a car stops short the instant it hits a wall, say, the driver may endure a split-second peak load of 100 G's. If the car has a collapsing hood—a common safety feature these days—the energy of those same 100 G's is released more gradually, reducing the peak force to maybe 10 G's—highly survivable.

The longer it takes the car to stop moving, the better—with one dangerous exception. To understand it, you need to understand what is happening to a body during a crash. Different types of tissue accelerate more quickly or slowly, depending on their mass. Bone accelerates faster than flesh. Your skull, in a lateral impact, leaves your cheeks and the tip of your nose behind. You can see this in a freeze-frame of a boxer's face* as he's punched in the side of the head. In a head-on, your frame gets moving first. It's hurled forward until it's stopped—by the shoulder belt or by the steering wheel—and then it rebounds backward. A fraction of a second later than your frame began moving forward, your heart and other organs depart. This

---

* And in the paper "Voluntary Tolerance of the Human to Impact Accelerations of the Head." Eleven subjects, at least one of them dressed in a suit and tie, received blows to the head with 9- and 13-pound pendulums. As the authors put it: "Considerable distortion of the face was observed as the bony structure of the head was accelerated away from the softer portions." We owe these men a debt of thanks. In the early investigations of head impact, a cadaver was of limited help. You couldn't ask him to count backward by sevens or name the president, and you'd never know what sort of headache he had.

means that as the heart is launched forward, it collides with the ribcage on its journey back the other way. Everything's moving forward and back at different rates, colliding with the chest walls and rebounding. And all of this is happening within a few milliseconds. So fast that *bouncing* and *rebounding* are the wrong words. Things are *vibrating* in there.

The big danger, Gohmert explains, is if one or more of those organs starts vibrating at its resonant frequency. This will serve to amplify the vibrations. When a singer hits a note that matches the resonant frequency of a wine glass, the glass starts to vibrate more and more energetically. If the note is sung loud enough and sustained for a long enough time, the glass will shake itself apart. Recall, if you are old like me, the Memorex ads with Ella Fitzgerald and the exploding wine glass. The same sort of thing can happen to an organ that hits its resonant frequency in a crash. It can shake itself off its moorings. And worse. "Essentially," said Gohmert, after repeated wheedling for specifics, "you're churned to death."

You may be wondering: Could Ella Fitzgerald explode your liver? She could not. Glass has a relatively high resonant frequency, up in the audible sound wave range. Body parts resonate down in the long, inaudible wavelength range called infrasound. A launching rocket, on the other hand, creates powerful infrasonic vibration. Could those sound waves shake apart your organs? NASA did testing on this back in the sixties, to be sure, as one infrasound expert told me, "that they didn't deliver jam to the moon."

Bolte's students are sliding F onto a stretcher and loading him into the back of a white van. He's traveling to the OSU Medical Center where he'll be scanned and X-rayed. The whole procedure will unfold exactly as it would with a live patient, right down to a forty-five-minute wait and a problem with the billing.

Gohmert's gaze rests on F. It is hard to read his look. Is he uncomfortable with having had to impact a human body? He turns to Bolte. This I didn't see coming. "Do you ever put 'em in the front seat and take 'em through the HOV lane?"

I RECALL AN IMAGE from early this morning. Two of Bolte's students, Hannah and Mike, are standing beside F, talking and laughing as they untangle the long, fine wires that trail from the strain gauges mounted on F's bones. Rather than seeming gruesome, the scene had a comfortable, familial feel, like a family stringing lights on the Christmas tree. I was struck by how at ease the students were. To them, the cadaver seemed to inhabit an in-between category of existence: less than a person, but more than a piece of tissue. F was still a "he," but not someone you needed to worry about hurting. Hannah, in particular, had a lovely way with him. While F lay in the CT scanner late that night, an automated recording commanded, "Hold your breath." "He's really good at that," she said. It was funny, but also a sideways acknowledgment of the unusual talents and abilities of the dead.

Not quite so at ease were the NASA team. Outside the context of the testing (and the carpool lane bit), they made very few references to him, and usually with the pronoun *it*. Getting permission to be here entailed months of emailing with a NASA public affairs officer and culminated in a flurry of tense phone calls upon my arrival this morning. Dead people make NASA uncomfortable. They don't use the word *cadaver* in their documents and publications, preferring the new euphemism *postmortem human subject* (or, yet more cagily, PMHS). In part, I'm guessing, it's because of the associations. Corpses in spaceships take them to places they'd rather not revisit: Challenger, Columbia, the Apollo 1 fire. And partly, they are unaccustomed to it. I have come across only one project that made use of human cadavers in the past twenty-five years of aeromedical research. In 1990, a human skull rode Space Shuttle Atlantis, kitted out with dosimeters, to help researchers determine how much radiation penetrates astronauts' heads in low Earth orbit. Worried that the astronauts would be unnerved by their decapitated crewmate, the researchers covered the bone with pinkish plastic molded to approximate a face. "The result was far more menacing than plain bone would have been," noted astronaut Mike Mullane.*

Back in the Apollo era, the agency's discomfort over using dead people in capsule impact studies appeared to transcend any discomfort they felt about using live ones. In 1965, NASA collaborated with the Air Force on a series of tests very similar to today's—but with human volunteers. Personnel from Holloman Air Force Base, seventy-nine in all, rode an ersatz Apollo space capsule seat on an impact sled while wearing helmets and other spacesuit components. The men endured 288 simulated splashdowns: upside down and right side up, backward, forward, sideways, at 45-degree angles. Peak forces were as high as 36 G's, more than twice as powerful as the 12 to 15 G's inflicted on Subject F today.

Colonel John Paul Stapp, a pioneer in human impact tolerance research, breezily summed up the project in a press release: "It might be said that at the cost of a few stiff necks, kinked backs,

* Here's another possible reason NASA avoids cadaver research: astronauts. "I floated into a sleep restraint and extended my arms through the armholes then ducked my head into the bag," wrote Mullane. "Pepe and Dave taped the skull on top . . . They silently floated the bag to the flight deck and maneuvered me behind John Casper, who was engaged at an instrument panel. When he turned to find the creature in his face with arms waving, it scared the bejesus out of him. Later, we clamped [it] on the toilet." If you read just one astronaut memoir in your life, make it Mullane's.

bruised elbows, and occasional profanity, the Apollo capsule has been made safe for the three astronauts who will have perils enough left over in the unknown hazards of the first flight to the moon."

I spoke to a man who rode Holloman's Daisy sled six times, in various positions, while wearing an Apollo helmet. Earl Cline is sixty-six now. His last ride was in 1966—25 G's. I asked Cline whether he'd suffered any lasting damage. He replied that he hadn't had any problems, but as the conversation went on, things began to emerge. To this day, he has pain in the shoulder that bore the brunt of a lateral impact. At the time of his discharge, he was found to have a torn heart valve and one eye that's "off a little bit."

Cline reserves his sympathy for the guy whose eardrum ruptured and the one who rode the Apollo seat upside down "with his rear end up in the air" and wound up with a ruptured stomach.

Cline expressed neither resentment nor regret, and has not pursued a disability claim. "I am very proud of the fact that I contributed. I like to think that when they went up in the Apollo missions their helmets didn't shatter or anything because I tested them." A subject named Tourville expressed a similar sentiment in a newspaper interview at the time of the Stapp "a few stiff necks" press release: "As long as I know this will save our Apollo astronauts from being hurt on their landings I don't mind losing sleep with a stiff back for a few days." Tourville took 25 G's and suffered a compression injury of the soft tissue surrounding three vertebrae.

Added motivation was provided by a generous hazardous-duty stipend. Bill Britz, a Holloman Air Force Base veterinarian, recalls being paid an extra $100 a month. Cline received $60 to $65 a month for riding the sleds a maximum of three times a week.

Given that his base pay was $72 at that time, it was a significant amount. "I lived like an officer," Cline told me, adding that there was a waiting list to become a Daisy sled volunteer. This was not the case over at Stanley Aviation, in Denver, which NASA had contracted to do some landing impact studies. Capsule mock-ups were hoisted aloft and then dropped onto surfaces of differing compressibility to see what sorts of injuries an astronaut might have to cope with should the capsule go off course and land not on water, but on dirt or gravel or the Winn-Dixie parking lot. There, Britz told me, the pay was only $25. "They got derelicts from Skid Row!" You would think that a news scandal involving underpaid indigents would be a scarier prospect for NASA than one involving cadavers, but things were different back then. The homeless were "derelicts" and "bums," and cadavers were people who rest on satin pillows.

THE FIRST AMERICAN to live through a space capsule landing mishap endured 3 G's more than the mission planners had anticipated. His capsule arced 42 miles higher than it was meant to and landed 442 miles off course. By the time rescue ships reached it, two and a half hours later, it had taken on 800 pounds of water and was partly submerged. With great trepidation, the hatch was opened. The space traveler was alive! Upon returning to base, he leapt into the waiting arms of Air Force Master Sergeant Ed Dittmer.

The astronaut was the three-year-old chimpanzee called Ham. (Dittmer was Ham's trainer.) Ham was more than just the first space capsule landing mishap, of course. He was the first American to ride a capsule into space and come back down alive. As such, he put a bit of a tarnish on the Mercury astronauts' con-

ONE FURRY STEP FOR MANKIND

The Strange Careers of Ham and Enos

The John P. Stapp Air and Space Park is made entirely of things that can hurt you. Eleven historical missiles are displayed amid plantings of spiny desert succulents. You walk along the gravel pathways, reading the little signs: PRICKLY PEAR, LITTLE JOE, CRIMSON HEDGEHOG. From the names alone, it is sometimes hard to know which is which. Is TURKS HEAD a cactus or an exploding munition? A similar sort of confusion can be found 25 yards down the hill, at the base of the flagpoles that mark the entrance to the park and the adjoining New Mexico Museum of Space History and International Space Hall of Fame. Flush to the pavement is a bronze grave marker that says, WORLD'S FIRST ASTROCHIMP HAM.*

---

* A comma would have been good. "Astrochimp Ham" is perilously suggestive of a cut of meat made from a dead research animal. It wouldn't be a first. In a stunning public relations lapse known as Project Barbecue, pigs who died on Air Force crash sleds in a 1952 test of seatbelt safety were served in the mess hall later that night.

The astrochimps were a knotty chimera. People weren't sure how to think of them. Chimps or astronauts? Research animals or national heroes? They're still not. Someone has left a basket of flowers on the grave, and someone else has left a plastic banana.

You can't blame people for being confused. The careers of Ham and Enos—the chimpanzees who, in 1961, flew the dress rehearsals for the first U.S. suborbital (January) and orbital (November) flights were in some respects not all that far off from the careers of Alan Shepard and John Glenn. The chimps and the two astronauts who followed them into space did not train together, but they could have. They spent time in the same altitude chambers and tried out weightlessness on board the same parabolic airplane flights, rode the same spinning centrifuges and vibration tables to get used to the noise and shudder and G's of liftoff. Come the big day, astrochimp and astronaut would suit up and ride out to the gantry in the same Airstream trailer.

For both species, piloting duties were light to nonexistent. Mercury capsules, as Ham's veterinarian Bill Britz says, "were not flying machines, they were bullets." Shoot them up, cue the parachutes, watch them come back down.* Speaking of both man and chimp, Britz said, "They were organisms placed on board." The science of the Mercury program was an extension of the V-2 and Aerobee and parabolic flights that led up to it. Aerospace biologists had established that humans can function for a few seconds without gravity. But what about an hour, a day,

---

* Piloting *could* be done by the astronauts, via directional thrusters, but it didn't need to be. The capsule could be flown on autopilot and operated from the ground in, to quote astronaut Mike Collins, "chimp mode."

a week? "People ask, *Why*?" says Britz of the era of the spacefaring chimp. "Mary, we just didn't know." What were the longer-term effects of space travel—not only of weightlessness, but of cosmic radiation? (High-energy atomic particles have been zinging through space at ferocious speeds since the Big Bang. Earth's magnetic field protects us by deflecting cosmic rays, but in space, these invisible bullets smash unimpeded through cells, causing mutations. It's serious enough that astronauts are classified as radiation workers.)

Just as the Alberts laid the groundwork for the Mercury fliers, Ham and Shepard and the rest would pave the way for the Gemini astronauts. And on it went. Gemini paving the way for Apollo. Six-month space station missions paving the way for the eventual long haul to Mars. Each space program along the way provides opportunities for planetary science, but in the grander scheme of space exploration, every program is fundamentally practice and prep for longer, farther trips to come.

Zero gravity still had NASA spooked. "The big bugaboo was weightlessness," said John Glenn in a 1967 Associated Press interview. "Many ophthalmologists thought the eye would change its shape and that this would change the vision, so that maybe the man in space would not be able to see at all." That is why, if you'd looked inside Glenn's capsule, you'd have seen a scaled-down version of the classic Snellen eye chart taped to the instrument panel. Glenn had been given instructions to read the chart every twenty minutes. A color blindness test and a device to measure astigmatism were also on board. I used to hear about Glenn's historic flight and think, "Man, what was that like—being the first NASA astronaut to orbit the Earth?" Now I know. It was like visiting the eye doctor.

An overabundance of gravity—the multiple G's of launch

and reentry—also had NASA concerned. An astronaut needed to be able to reach the instrument panel in case something went wrong. If his outstretched arm weighed 70 pounds instead of 9, would he have the strength to raise it? This is why Ham (and later Enos) spent weeks learning a routine that would have them reaching over to an instrument panel and pulling levers throughout their flights. The lever-pulling also let researchers keep track of any cognitive ebbs during the chimps' flights. They wanted to be sure that zero gravity, combined with some yet-to-be-discovered X factor, wouldn't disorient a space flyer or slow his reaction time.

Given that the Mercury fliers were gold-standard, swinging-dick military test pilots, the concern did not sit well. These men hadn't been in space, but they'd spent enough time on the doorstep to feel confident they'd be fine. As test pilots, they'd endured G forces during climbs and pullouts that were higher and more sustained than any they'd have to deal with on a Mercury flight. They didn't worry about their abilities; they worried, if anything, about their ride. As of two months before launch, the guidance system of the Redstone rocket that would carry Shepard's capsule into space had been misbehaving, and there were seven last-minute modifications to the hardware that hadn't been tested in flight. That's another reason NASA sent chimpanzees up first. (They would come to regret the caution. Three weeks before Alan Shepard launched, cosmonaut Yuri Gagarin became the first man in space.)

Ham's flight implied—in a widely publicized manner—that the astronaut, America's hero, was no more than a glorified chimp. "To be preceded by a chimpanzee was just a blow to their ego," Bill Britz told me. The astronauts would surely have preferred another quiet dummy launch. In the months prior to Ham's flight, a capsule was launched carrying a "crewman simulator"* that "breathed," consuming oxygen and producing carbon dioxide to test the cabin sensors. The same insinuations could be made about a man whose job could be done by a dummy, but the press didn't cover dummy flights the way they covered chimp flights. The banana pellet dispenser was gone when Shepard and Glenn climbed on board, but the stigma remained. As fighter jock Chuck Yeager, the rightest of stuff, famously put it, "I wouldn't want to have to sweep monkey shit off the seat before I climbed into the capsule."

Though Ham and Enos and their alternates lived and trained in trailers alongside the astronauts' live-work quarters at Kennedy Space Center's famous Hangar S, Britz says he can't recall talking to Alan Shepard more than once or twice. "We didn't mingle much." Enos's veterinarian Jerry Fineg agrees: "They didn't want to recognize the fact that we were there." Chimp jokes were poorly received. Britz told me a story about a placard posted on the wall of the van that both astrochimp and astronaut rode to the launch

---

* The simulated astronaut is a tradition dating all the way back to the Sputnik era, when the Soviets flew test runs with a mannequin they called Ivan Ivanovich and, sometimes, recordings to test voice transmissions. A tape of a person singing was originally proposed, so as to make clear to Western listening posts that it wasn't a spy. Someone pointed out that this would generate rumors of a cosmonaut spy gone mad. The recording was switched to choral voices, as even the most gullible Western intelligence man knew you couldn't fit a choir in a Korabl-Sputnik satellite. A voice reading a Russian soup recipe was thrown in for good measure. The simulated astronaut named Enos orbited with a voice-check tape recording that said, "Cap com, this astro is. Am on the window and the view is great . . . ," prompting President Kennedy to announce to the world, "The chimp took off at 10:08. He reported that everything is perfect and working well." No doubt generating KGB rumors of a U.S. president gone mad.

pad. "They had Alan Shepard's trajectory plotted on [it]. We very carefully plotted Ham's trajectory higher and farther." (Owing to a malfunction, Ham flew 42 miles higher than planned.) "I'm telling you, it really pissed some people off. That thing disappeared in a minute." Mercury launch pad director Guenter Wendt once reprimanded Shepard by threatening to replace him with one of those guys who works for bananas. Shepard, the story goes, threw an ashtray at his head.

Chimp humor was less nettlesome for John Glenn than it had been for Alan Shepard, because Enos wasn't the media sensation Ham had been. At the time Ham flew, a pair of Soviet dogs, Belka and Strelka, had already returned alive from orbiting Earth, and the press was impatient for a U.S. milestone in space. When Ham splashed down alive, they presented him less as a research animal than as a sort of short, hairy astronaut. The chimp appeared on the cover of *Life* magazine in his mesh flight suit* beside the headline, "A Confident 'Ham.' Back from Space." The public sopped it up. Letters and flowers and gifts addressed to Ham began arriving at the chimpanzee colony at Holloman Air Force Base, where Ham returned after his flight. People sent their copies of *Life* with requests for Ham's "autograph." Holloman staff gamely complied, the little hand pressed on inkpads over and over, so many times that a copy of *Life* "autographed" by Ham fetches just $4 on eBay. (And is possibly a fake: Fear-

---

* Ham and Enos traveled in pressurized compartments and thus didn't need pressurized spacesuits and helmets. Nonetheless, some prototype chimp suits had been developed, including the "SPCA Suit"—certified humane by the Society for the Prevention of Cruelty to Animals. "To prove that a suit was safe for a man, we were going to test it on a chimp, but to prove the suit was safe for a chimp, we had to test it on a man," *U.S. Spacesuits* coauthor Joe McMann said in an email. "That was a mind boggler."

ing they'd "wear him out," the staff, Britz told me, "just put any chimp's hand on it after a while.")

Newspaper databases typically have about five times as many Ham stories as Enos stories. "Enos didn't have the charisma, and he wasn't first," says Fineg. Thus, John Glenn's glory was little diluted by his simian predecessor. Also, Glenn managed to deflect the unkind comparisons by making the jokes himself. He told a congressional audience about the humbling experience of having been asked by President Kennedy's young daughter Caroline, while her father stood by, "Where's the monkey?"*

Enos was as unpopular as Ham was beloved. In news accounts, you could tell Fineg had applied himself to the task of finding positive ways to describe Enos. Rather than "obstinate" and "ornery," terms he currently uses, Fineg referred to Enos as a "quiet, taciturn, pillar of the community type."

"He was a mean one," Fineg recalled when we spoke. Staff nicknamed him Enos the Penis. "Because he was just a son of a gun."

"Meaning he was a dick."

"Yeah."

The nickname Enos the Penis is mentioned in the book *Animals in Space*, but the authors have an altogether different account

---

* A lasting fixation for young Caroline. Three months earlier, around the time of Enos's flight, Jackie Kennedy rented a monkey for her daughter's first birthday party in the White House, an event widely covered by the wire services at the time. In addition to a live monkey, the party featured "jelly sandwiches," whistles, tricycles going "up and down the ground floor of the White House," and, hopefully, sedatives for Jackie. Caroline no doubt wanted her very own space chimp. It was a reasonable expectation, given that Nikita Khrushchev had presented her mom with one of the puppies of space dog Strelka. The puppy was a gift, but also a nose-thumb: Strelka had beaten Enos into orbit by over a year. According to *Animals in Space*, White House staff had the pup searched and X-rayed "to check for bugs or a possible doomsday device."

of its genesis. They write that "Enos the Penis" derived from the chimp's fondness for masturbating, and that NASA had inserted a balloon catheter in his penis during his orbit in part to discourage the habit. (Both Ham and Enos were to be filmed during their flights.) When the lever system malfunctioned, delivering shocks rather than banana pellets for correct responses, a frustrated Enos had yanked out the catheter and "began fondling himself in front of the camera." Or so the story went.

I spent a few breathless days searching government archives for the X-rated Enos footage. I found footage of Ham in flight and Enos being readied for flight, but none of Enos inside the capsule pulling levers—his own or NASA's. I contacted Fineg again.

"I don't know where that came from," he said. "I worked with Enos for a number of years, and never saw him do anything like that. His name was the result of his demeanor."

"So the catheter didn't have anything to do with keeping him from touching himself?" I don't usually go in for euphemisms, but Fineg is a man who says "behind," as in "I have a picture where he bit me in the behind." The catheter, it turns out, was in the chimp's femoral artery (to monitor blood pressure), not his urethra.

Still mildly unconvinced, I called Fineg's colleague Bill Britz, who had been Ham's vet but also worked with Enos.

"Naw," said Britz. "I mean, most male chimps play with themselves. But he couldn't even get to it." Britz explained that the couch inside the capsule was designed with a barrier to keep the chimp from reaching down below the waist and pulling out the arterial catheter during the flight. Britz agreed with Fineg: Enos had no such reputation.

I contacted Chris Dubbs, one of the authors of *Animals in Space,* to find out where the story had come from. He forwarded an article his coauthor had found on the Web site of a Dr. Mohammad Al-Ubaydii. The Al-Ubaydii rendition included an arresting new detail:

"During the ensuing press conference, Enos began by pulling his nappy down. NASA's people were horrified of what might follow. Fortunately Enos had more class than this, and restrained himself."

Dr. Al-Ubaydii, replying to an email, said he'd come upon the story in the 2007 book *Space Race.* In this version, Enos is less restrained: "As he pulled down his trousers, cameras clicked, flashing like diamonds, ensuring that Enos['s] name would live in memory as much for his hobby as for his aeronautical achievements." Inquiries to the author produced no reply, but a Google Books search unearthed another reference, this one in *Dark Side of the Moon,* published in 2006. "The next day at his post-flight press conference, he horrified his NASA handlers when he ripped off his diaper and started to fondle himself." *Dark Side* cites yet another book on the Apollo race: James Schefter's 1999 *The Race.*

"[Enos] would pull down his diaper in the middle of a training exercise and begin to masturbate. His handlers and medics figured that he'd stop if they inserted a catheter to drain off urine instead of using a condomlike device attached to a tube. It didn't work. . . . They devised an advanced catheter with a small inflatable balloon to prevent its easy removal." In those few lines, Schefter establishes himself as, in the words of one reviewer, a writer who "does not let facts get in the way of a good story." The condom–tube device sounds like the urine collection device designed for Mercury astronauts to use during spaceflight. It was never used on chimps. And it is hard to imagine anyone going through the significant risk and hassle of catheterizing a chimpanzee just to keep him from playing with himself during training sessions. As for the balloon catheter, it was patented in 1963—two years *after* Enos's flight—as a tool to remove blood clots, not to discourage chimpanzee masturbation. *The Race* has no sources or bibliography, and Schefter died in 2001.

What's interesting is that Schefter never says Enos was masturbating during his spaceflight. He merely states that he pulled his catheter out. Nor did he claim that Enos fondled himself at the postflight press conference (which took place uneventfully at Kindley Air Force Base in Bermuda, not far from where Enos's capsule was recovered). Schefter's scene takes place back at Kennedy Space Center, not at a press conference but in front of a few reporters and NASA officials, as Enos descends the steps of the plane that brought him back from Bermuda. And he merely pulls his diaper down.

The story, as stories will, grew and mutated with each retelling, until Enos was having the world's first orbital orgasm and then coming back down and brazenly masturbating in front of a sea of clicking cameras and exploding flashbulbs.

Here is the opening of the story the AP reporter filed after attending the infamous postsplashdown press conference in Bermuda. "Holding his first public audience since returning from outer space, the Holloman Air Force–trained chimpanaut refused Thursday to do even a cartwheel for newsmen at his press conference. 'He's really quite a cool guy and not the performing type at all,' said Captain Jerry Fineg."

Enos, your name is cleared.

A BLOW-DRYER wind has knocked over the flowers on Ham's grave. I'm out here squinting in the noon sun, eating a sandwich and thawing out after a morning in the museum's aggressively air-conditioned archives. Now I know the story behind the plaque. The same confusion that surrounded Ham while he was alive continued when he died. The International Space Hall of Fame was bombarded (their wording) by inquiries from the media and the public about the fate of his remains. It was something of a quandary. What's appropriate protocol for a dead space chimp? Memorial service or incinerator?

The Air Force's position was made clear in a draft of a letter by a Colonel William Cowan: Ham was a historical artifact. Cowan, repeatedly referring to Ham's remains as "the carcass," recommended that following the necropsy (the animal version of an autopsy), the skeleton be removed from the body and cleaned of flesh in the Smithsonian's dermestid beetle colony and then sent to the Armed Forces Institute of Pathology archives.

Ham's hide had already been removed, in case the Smithsonian wished to prepare a taxidermied specimen. This seemed like a bad idea to me. I saw a photograph of Ham taken ten years after his flight. He had gained more than a hundred pounds over the course of his retirement and lost some of his teeth. Others protruded at unfetching angles. He was unrecognizable as the flight-suited, pink-faced youngster from the *Life* cover. He looked like Ernest Borgnine.

But no one asked my opinion. The Smithsonian announced plans to stuff Ham and add him to "the indoor Ham exhibit" at the International Space Hall of Fame, an exhibit that consisted at that time of "a photo of Ham." The public went bonkers. The archives has a few of the letters. "Gentlemen: Ham is a national hero and not a thing. . . . Do you propose to stuff John Glenn as well?" "A chimpanzee is *not* a stuffed pepper." Et cetera. The *Washington Post,* under the inevitable "The Wrong Stuff" headline, took the nation's indignation a step further in an op-ed that insinuated Communist proclivities on the part of the Smithsonian. "The only national heroes we can think of who are stuffed and on permanent display are V. I. Lenin and Mao Tse-tung." (In keeping with the Communist proclivity for stuffing heroes, Soviet space dogs Belka and Strelka stand side by side in glass cases in Moscow's Memorial Museum of Cosmonautics, faces raised as though staring at the heavens or anticipating a treat.)

A follow-up announcement was quickly drafted. Ham would

not be stuffed. He would be given "a hero's burial" in a small plot in front of the Hall of Fame flagpoles, "similar to the final resting place of Smokey the Bear."* What remained of Ham following a necropsy, a skeleton extraction, and the removal of his hide is difficult to imagine. Whatever it is, one has to assume, is what's down there under the flowers.

The museum now had to come up with a suitable memorial service. They needed a respected public figure willing to say a few words about Ham's contributions to manned space exploration in the United States. Clearly heat-struck, their public relations person sent off a letter to notable Ham detractor Alan Shepard. The letter pointed out that Shepard would enjoy "national attention from all areas of the media." As though Alan Shepard, the first American man in space, wanted or needed media attention. In particular, at an event that would yet again have him sharing the spotlight with a chimp. The letter-writer acknowledged the "jokes and sometimes 'unfunny' humor about the situation." The quotation marks were an ill-advised touch, seeming to suggest that the letter-writer herself found the jokes funny.

A reply arrived on letterhead from the Texas-based Coors distributorship where Shepard served as president, thanking the museum for the "thoughtful invitation" and expressing regrets. The letter was typed by Shepard's secretary, initials JC. There was no signature. Undiscouraged, the Hall of Fame public relations staff next went after John Glenn, by this time not just an astronaut

---

* Curiously, also situated in New Mexico. The Smokey plot does not contain the remains of the Forest Service mascot, who is a cartoon, but of a black bear cub burned in a New Mexico fire and named after the mascot. Confusion surrounds the official mascot name, which is Smokey Bear, not Smokey the Bear. Just as the official slogan of New Mexico is Land of Enchantment, not Land of Pants-Wearing-Animal Memorials.

but a senator and a presidential candidate. Glenn politely declined, citing previous commitments.

A brief news story on the ceremony ran in the *Albuquerque Journal*. A photograph accompanying the article showed a loose crowd of maybe forty people standing around the flagpole area. "Colonel Stapp made a short speech and members of Girl Scout Troop 34 of Alamogordo laid a wreath on a small memorial plaque." Stapp ran the crash sled research program at Holloman Air Force Base. In both aerospace and automotive safety studies, Holloman chimps were regularly used in impacts deemed too hazardous for airmen. Which made Stapp both an appropriate and inappropriate choice. He was intimately familiar with the heroic sacrifices of man's closest cousin; he'd signed the paperwork on most of those sacrifices himself. The tribute was respectful, if short on sentiment*—one of those rare eulogies to incorporate numeric G force figures.

Enos had no memorial. A log book of Holloman chimp acquisitions† includes the note "remains at Smithsonian," though no one there seems to know where he ended up. *Animals in Space* author Chris Dubbs spoke to someone whose mother had dissected Enos's eyes to study the effects of cosmic radiation, but the

---

* Not that Stapp was unsentimental. The colonel composed sonnets and love poems for his wife Lillian, a ballerina with the American Ballet Theatre. They're included in a collection of Stapp's verse, on sale for $5 in the New Mexico Museum of Space History gift shop. Stapp didn't read from his oeuvre at Ham's service, though one line in particular would have fit the occasion: "If chimpanzees could talk, we would soon wish they wouldn't."

† Ham is entered twice, initially as "Chang," and later as "Ham" (an acronym of Holloman Aeromedical). Once the animal had been chosen as a finalist to fly, government officials rethought the name, worrying that an ape named Chang might offend the Chinese. To be on the safe side, chimps were thereafter named for Holloman staff or, in the case of Double Ugly, Miss Priss, Big Mean, and Big Ears, themselves.

man knew nothing about the rest of the chimp. This suggests that the body was parceled out for research. Which is the usual and appropriate fate of a research subject.

For better or worse, that's what Ham and Enos were. They played a vital role in the country's space efforts, but I would not use the term "heroes." For the simple reason that no bravery was involved in what they did. A courageous feat is one undertaken with an understanding of the dangers involved. As far as Ham knew, January 31, 1961, was just another strange day in the little metal room. Alan Shepard may not have been using the expertise of a test pilot, but he was certainly using the guts. He let himself be strapped in a canister on the nose of a missile and blasted into space: an insanely dangerous feat undertaken by, at that point, only one other man.

The decision to put a chimpanzee in space before an astronaut was not, in either instance, an easy one. NASA had to weigh concern for the Mercury crew and lack of confidence in the hardware against the enormous pressure to best the Soviet Union. The early days of the Apollo program would be plagued with the same mixture of urgency and caution. Having watched the USSR rack up space firsts—first man-made satellite, first orbit of a live animal (Laika), first recovery of live animals (Belka and Strelka) from orbit, first man in space and in orbit, first spacewalk—the United States was ever more determined to reach the moon first. NASA was working furiously on President Kennedy's publicly announced time line: By the end of the 1960s, America would put a man on the moon. Or anyway, something pretty close.

First U.S. Flag on Moon May Be Planted by Chimp

BETWEEN MAY 1962 and November 1963, veteran Associated Press reporter Harold R. Williams filed four stories based on visits to a new chimp facility at Holloman Aeromedical Research Laboratory. "Chimp College," as he called it, was a million-dollar expansion of the grotty-looking facilities where Ham, Enos, and other chimps had lived and trained for the Mercury missions. It featured a staff of twenty-six, brand-new "dorms" with an outside run attached to each cage, a surgical suite, a kitchen, and a curriculum of "new, complicated and secret" tasks. Williams's series ran in dozens of U.S. newspapers under various headlines like the one above, almost all of them highlighting the possibility of a lunar mission: "First from U.S. to Moon? Chimponauts* Hard at Work on Secret Space Program." "Holloman Monk May Be First on Moon." "Space Chimps' College Grad May Hit Moon."

Williams described college "Ph.D." Bobby Joe as he sat at an instrument panel mock-up, effortlessly maneuvering a joystick to keep a crosshair centered inside a circle. "There is no question about it," said Williams's guide, a Major Herbert Reynolds, who would go on to become president of Baylor College of Medicine. "He could guide a space vehicle into space and bring it back." On a different visit, Williams peered through the window of a "simulated space vehicle" at a chimp named Glenda. Glenda had been inside for three days, sleeping and working on the same shifts an astronaut would have. She had two days left to go.

Five days is what it took the Apollo 11 astronauts to reach the moon and plant the American flag. Was it true? Had NASA and the Air Force been planning to beat the Soviets to the moon by sending a trained chimpanzee on a one-way mission? A round trip was certainly out of the question. Lifting off from the moon and docking

---

* Holloman moved away from this term after receiving letters from irritated etymologists. The suffix "naut" comes from the Greek and Latin words for ships and sailing. *Astronaut* suggests "a sailor in space." *Chimponaut* suggests "a chimpanzee in sailor pants."

with an orbiter was beyond the capabilities of an ape. But a straightforward moon shot and capsule touchdown could be managed from the ground, just as unmanned rovers are landed remotely today.

The trickiest part would be finessing the public relations debacle of a dead chimpanzee hero. Best not to take a cue from the Soviet playbook. In November 1957, a mellow and patient Moscow street dog* named Laika, traveling suitless in a pressurized capsule, became the first living creature to orbit the home planet. Alas, there was no plan or means to bring her safely back down. For over a week, Soviet officials were mum on the topic, refusing to say whether Laika was still alive. They ignored inquiries from media and animal rights groups, until the clamor and outrage had all but eclipsed the glories of their achievement. Finally, nine days after the launch, Radio Moscow confirmed that Laika was dead. The particulars were left to speculation. In 1993, Laika's trainer Oleg Gazenko told one of the authors of *Animals in Space* that she'd perished when a malfunction caused her capsule to overheat, just four hours into her flight.

Perhaps less scandalous to send a willing human. In 1962—the same year that Williams filed his Chimp College pieces—a story ran in a Sunday newspaper supplement called *This Week* suggesting that the USSR was considering sending a cosmonaut on a one-way lunar landing mission. That same year, according to space historian Dave Dooling, *Missiles and Rockets, Aviation Week & Space*

---

* According to space historian Asif Siddiqi, the Soviets preferred to train dogs for space travel, because apes were too excitable, too prone to catching colds, and "more difficult to dress." And because Soviet space program bigwig Sergei Korolev loved dogs. Both the United States and the Soviet Union built a Tomb of the Unknown Soldier, but only Russia has a Tomb of the Unknown Dog (outside St. Petersburg), honoring the contributions of canine research subjects.

Technology, and *Aerospace Engineering* all detailed a similar mission proposal making the rounds at NASA. The "one-way, one-man" lunar expedition was the brainchild of a pair of Bell Aerosystems engineers, John M. Cord and Leonard M. Seale. "It would be cheaper, faster, and perhaps the only way to beat the Russians," Cord is quoted as saying. Dooling points out that intelligence data gathered at that time suggested that the Soviets would be capable of landing a craft on the moon as early as 1965. (The United States landed on the moon in 1969.)

Neither the Soviet nor the American version proposed leaving the sad spaceman to die on the moon. Someone would come pick him up in one to three years—just as soon as they figured out how to do it and built the hardware. A total of nine launches would follow his own, delivering a living module, communications module and equipment, construction equipment to build the modules, plus the 9,910 pounds of food, water, and oxygen he was projected to consume while waiting around for his ride.

And who would agree to go? "It is sincerely believed," wrote Cord and Seale, "that capable and qualified people could be found to volunteer for the mission even if the return possibilities were nil." I believe it. There are astronauts today who happily would sign on for a one-way mission to Mars. This scenario holds no eventual return trip. Rather, the crew would live out the rest of their lives with help from unmanned resupply landers. "I've spent my life training to go into space," astronaut Bonnie Dunbar told *New Yorker* writer Jerome Groopman. "If my life ends on a Mars mission, that's not a bad way to go." Valentina Tereshkova, the first woman in space, said in a 2007 interview that reaching Mars was the dream of the early cosmonauts and that she would love, at seventy-two, to realize that dream: "I am ready to fly without coming back." Though years or decades of resupply launches might not be cheaper or easier than figuring out the technology to make fuel for the ascent engines out of Martian resources. Or putting fuel and hardware for the return trip onto those unmanned landers, instead of survival supplies.

Dooling thinks it unlikely that anyone at NASA gave serious thought to Cord and Seale's one-way moon mission. But it does lend credence to the possibility that the aerospace community had —however fleetingly—considered launching a one-way chimped mission.

I went back and reread Williams's AP stories. Outside of the headlines, there were no specific references to a lunar mission. Were the newspaper* editors taking liberties to make the story more provocative? I needed another source. Major Reynolds is dead. Jerry Fineg had left Holloman by 1962. Both he and Britz said they didn't recall hearing anything about it, though Britz recalled seeing rhesus monkeys at Brooks Air Force Base, near San Antonio, being taught to operate a joystick. "They were trying to see if they could actually fly," he told me in an email. "They were good!" Britz didn't know what the ultimate goal of the project had been. I do know chimpanzees were being trained for space-related tasks at Brooks as late as 1964, because I found a paper that referred to a chimp injured in the spacecraft simulator when the foot plates malfunctioned and delivered more than the customary "small but annoying" electrical shock.

Air Force historian Rudy Purificato is at work on a history of

---

* These were not great papers. Headlines proclaimed absurdities like "Black Label Was Elected a Fine Beer" and "Science Cures Piles!"—advertisements misleadingly typeset to look like news. Not to mention the very confusing "Thieves Get Ham." In what I first took to be an astrochimp kidnapping plot, two men pried open the rear door of a supermarket and made off with a dozen three-pound canned Rath Blackhawk hams and a half-dozen canned Wilson (clearly the inferior ham) half-pounders.

Wright-Patterson Air Force Base, the other hotbed of aerospace medicine research in the sixties. I sent him a note. "There could very well have been actual plans to send a chimp to the moon," he replied. He added that most of the primate research was still classified, and in that case Fineg and Britz (and Purificato) couldn't talk about what they knew. So who would have told the AP reporter? He had probably, Purificato said, benefited from a "slip-up" by someone he interviewed.

Holloman Air Force Base is a ten-minute drive from the New Mexico Museum of Space History. Perhaps the base archives could provide some answers. The curator here at the New Mexico museum, George House, gave me a phone number to try. The staff played hot potato with my call until someone could locate the Person in Charge of Lying to the Press. The PCLP said that the room that houses the base archives is locked. And that only the curator would have a key. And that Holloman currently has no curator. Evidently the new curator's first task would be to *find a way to open the archives.* Now I was sure of it: the chimp-to-the-moon files were locked up in there along with the Enos in-flight sex tapes and pictures of Colonel Stapp in a tutu. Paranoia is a way of life here in Alamogordo, home of the first atomic bomb test and not far from Roswell and Area 51, the secretive Air Force experimental aircraft proving ground/UFO hub. House said that emails containing the word *primate*, including some from me, mysteriously disappear en route to his computer. But House didn't think it had anything to do with secret chimp moon missions. He said it had to do with a lawsuit filed by People for the Ethical Treatment of Animals. The suit isn't against the Air Force per se, but rather the facility they'd contracted to take over the care—"care" being a rather gross overstatement—of the chimpanzee colony in the 1970s, when the Air Force no longer had use for them. Oh.

I went back out to the missile garden and paged through my

photocopies again. I noticed something I'd overlooked. One of the articles said that before being taken out of the capsule, the chimp Glenda "had to re-enter through the jarring forces of earth's atmosphere." That meant Glenda's simulated mission was round-trip, not one-way.

I'm guessing that Glenda was a simulated Gemini astronaut. (The Gemini space program, 1965 to 1966, was the precursor to the Apollo program's lunar missions.) From 1964 to early 1966, "Chimp College" primates were called on to provide answers to questions like, What will happen to an astronaut if his pressure suit tears while he's outside the capsule? "Previously," said the AP reporter who covered a series of chimpanzee-crewed EVA simulations designed to answer that question, "scientists believed direct exposure to space vacuum would result in death, with the blood boiling and the lack of atmospheric pressure possibly leading to the body expanding and even bursting."* Yet another reason Holloman can't get their archives door open.

That the prospect of a chimpanzee-piloted lunar mission was taken seriously enough to be printed as news demonstrates how political the Apollo space program had been. The goal? Pure and

---

* Contrary to popular lore, an astronaut's blood does not boil if his spacesuit tears or his craft depressurizes. And though he would swell, he would not burst. The body functions as a sort of pressure suit for the blood, keeping dissolved gases in their liquid state. Only body fluids directly exposed to a vacuum actually boil. (As happened to a 1965 NASA test subject in a leaky spacesuit in an altitude chamber. The last thing he recalled before losing consciousness was the sensation of his saliva bubbling on his tongue.) Also, current EVA suits are designed to compensate for tears or leaks by blasting in air at far greater pressure. Bottom line: Provided he has an oxygen supply, an astronaut in a spacecraft depressurization has about two minutes to figure out what's wrong and set it right. Beyond that he's in trouble. This is known from experiments in vacuum chambers that would, if you knew the details, make your blood boil.

simple: Land something before they do. Science on the first lunar surface missions was something of an afterthought: *Pick up some rocks while you're there, okay?* The first geologist wouldn't set foot on the moon until Apollo 17, six missions later.

The Cold War has ended, and the goals of space exploration are ostensibly grounded in science. There are those who argue that the science is more effectively—or cost-effectively, at least—carried out by robotic landers. And that the main reason to employ humans in space exploration and planetary science is to maintain the public's interest and support. As the saying goes, "No bucks without Buck Rogers."

Others disagree. "If your goal is to answer very specific questions like, How hard are the rocks on the surface of Mars? a robot is perfect. If your questions are bigger, like, What is the history of Mars? well, that's a hell of a lot of robots," says Ralph Harvey, a planetary geologist who has helped plan research expeditions on the moon. "But it could be only one or two human beings. Because human beings have this amazing tool called intuition, where they've built up a catalogue of experiences and they can draw on it instantaneously and spend one minute looking at a scene—whether it's on Mars or at a crime scene—and know what happened here."

For the past twenty-three years, Harvey has overseen the Antarctic Search for Meteorites, so he knows a great deal about doing geology under extremely harsh conditions. When we spoke, he had just returned from NASA's Goddard Space Flight Center, where he was helping plan a lunar traverse scheduled to take place sometime around 2025.*

Why does it take *fifteen years* to plan an outing on the moon? You'll see.

---

* Or not at all, if the 2010 NASA budget passes as is.

9

NEXT GAS: 200,000 MILES

Planning a Moon Expedition Is Tough, but Not as Tough as Planning a Simulated One

Once upon a time, astronauts tooled around the moon in an open two-seat electric buggy. It was the sort of thing one might see on a golf course or at one of those big Miami delis whose elderly patrons appreciate a lift to and from the parking lot. It gave lunar exploration in the seventies a relaxed, retirement-community feel. That's gone now. NASA's new rover prototypes more resemble a futuristic camper van. The entire cab is pressurized, which is good, because that means the astronauts can take off their bulky, uncomfortable white bubble-head EVA suits. The NASA shorthand for a pressurized interior is "a shirtsleeve environment," which makes me picture astronauts in polo shirts and no pants. If NASA ever builds an outpost on the moon,* astronauts will be undertaking rover traverses of unprecedented length and complexity. Teams of explorers will head out in two vehicles that rendezvous daily, finally returning to the base after two weeks on the roll. The new rovers sleep two and are equipped with a food warmer, a toilet with "privacy curtain," and cup holders (two).

Before actual prototypes of the pressurized rovers are tested in analog settings—earthly terrain that resembles the moon's surface—NASA is undertaking some rough cuts. These are two-day "excerpts" of fourteen-day traverses using similarly sized Earth vehicles. Simulated traverses help NASA get a hands-on sense of "performance and productivity"—how much gets done, how long things take, what works and what doesn't. This summer, the Small Pressurized Rover† simulator is an orange Humvee that lives at the HMP Research Station on Devon Island in Canada's High Arctic. (HMP stands for Haughton-Mars Project; Devon

---

* Up until Obama's first NASA budget appeared, in February 2010, the moon base was slated to be built sometime in the 2020s. That program (Constellation) has been cut, and now we're headed to a near-Earth asteroid and on to Mars. Then again, Congress has yet to approve the budget plan, so it's hard to know for sure, at the time of this writing, just where we'll end up hauling our rovers next.

† Six months after our traverse, NASA, recognizing a public relations opportunity, will change the name Small Pressurized Rover to Lunar Electric Rover. It was originally called the Flexible Roving Expedition Device, or FRED, until NASA Headquarters nixed it. They nixed it for the same reason they took the word Excursion out of the Apollo Lunar Excursion Module—it sounded frivolous. A larger mobile lunar habitat prototype called the All-Terrain Hex-Legged Extra-Terrestrial Explorer (ATHLETE) recently squeaked past the NASA fun censor. Whoever he is, he's very thorough. I skimmed the entire 53-page NASA acronym list and failed to find anything amusing. (Business Manager came closest.)

Island also resembles parts of Mars, and simulated Martian traverses have also taken place up here.)

In short, Devon Island is as close to the moon as you can come without a rocket. Twelve-mile-wide Haughton Crater is a ringer for the moon's Shackleton Crater, upon whose rim NASA had, since 2004, been planning to establish a base. Craters are formed by strikes from meteoroids* hurtling through space at somewhere in the neighborhood of 100,000 miles per hour; with no atmospheric friction to slow them down and burn them up, as happens above Earth, even tiny ones blast holes in the moon's surface. A pebble strike can open a crater a few feet across. Planetary scientists are fond of meteorites because they're natural excavators, yielding access to geological material from past eras that is normally costly and difficult to get to.

Devon Island is also, like the moon or Mars, extremely inconvenient. It's thousands of miles from the things one needs for a geology expedition. Devon is uninhabited: no electricity, no cell coverage, no port or airport or supplies. That is part of the draw. Doing science here is a lesson in extreme planning. A moon or Mars analog, rather than the orb itself, is the place to figure out that, say, three people might be a better size for an exploration party than two. Or that it takes twice as long as the mission planners thought to drive a rover over a block field or twice as much oxygen to climb the loose scree on the slope of a crater. As some-

---

* A meteoroid is a bit of debris, usually planetary, hurtling through the solar system. If it's bigger than a boulder, then it's an asteroid. If any part of a meteoroid makes it to Earth intact rather than burning up as it barrels through Earth's atmosphere, then it's a meteorite. A meteoroid's visible path through the atmosphere is a meteor. An astronaut struck by a meteoroid is a goner. A meteoroid the size of a tomato seed can pierce a spacesuit.

one at yesterday's pretraverse planning meeting said, "This is the place to make mistakes."

LIKE THE MOON, Devon Island doesn't get interesting until you start to get close. Out the window of a low-flying Twin Otter, ground that had appeared on satellite images to be dirt, straight no chaser, reveals itself to be riverine windings of tan, gray, gold, cream, rust. Polar meltwater has carved, scoured, and tinted the ground in such a way that you feel as though you're flying over an expanse of Italian marbled paper.

Set out on foot, and you soon see why planetary geologists make their way to the top of the Earth to visit this place. There are other places where meteorites have gouged out craters the size of Haughton, but most are covered with forest or mall. The High Arctic is landscape at its most elemental: earth and sky. Radiating out from the center of Haughton Crater is an "ejecta blanket" of the same kind you find around lunar craters. When a meteoroid slams a fellow celestial body, the energy of the impact simultaneously smashes and renders molten the rock beneath. The resulting magmalike rock stew is blasted away from the impact, lands, and cools into a sort of nougat, called impact breccia (pronounced as though it were an Italian delicacy). And then sits for 39 million years until some guy in hiking boots and a space helmet comes along and picks it up.

Today there are two guys in helmets. In the driver's seat of the Small Pressurized Rover simulator is planetary scientist and Haughton-Mars project director Pascal Lee. With support from NASA, the SETI Institute, the Mars Institute, and other partners, Lee established the HMP Research Station at Haughton Crater in 1997. Riding shotgun is Andrew Abercromby, of NASA's EVA Physiology Systems and Performance Project. Abercromby has blond, freckled good looks that are rescued from Buzz Lightyear all-American wholesomeness by a curious silver-dollar-sized circle of white hair and a Fyfe accent. Squeezed between Lee and Abercromby is HMP intern Jonathan Nelson and Lee's ubiquitous canine pal Ping Pong. Three all-terrain vehicles (ATVs) follow along behind the Humvee, carrying camp mechanic Jesse Weaver, spacesuit engineer Tom Chase, and me. Together we six are Small Pressurized Rover Alpha, or as "ground control" calls us, SPR-Alpha. Out on a different route, scheduled to rendezvous with us at the end of the day, are the men and women of SPR-Bravo.

We're driving slowly, keeping to the projected 6-miles-per-hour average of the actual rover. The low, gravelly hills are more uniformly grey here than elsewhere on the island. The scenery looks a lot like the moon's Taurus-Littrow Valley, where Apollo 17 astronauts explored by rover in 1972. Tooling along this barren terrain in a bulbous, visored ATV helmet, I find it easy, if embarrassing, to pretend I'm on the moon. Lee's evident excitement over the excursion—"Can you believe I get paid for this, barely?"—has become easier for me to understand. The place has made geeks of all of us.

Except our mechanic. Weaver never looks around to admire the scenery. I do, almost constantly. Yesterday, I came within inches of slamming the back of the ATV in front of me. Lunar scenery was a potentially dangerous distraction during Apollo landings. Concerned mission planners built gawp time into the minute-by-minute schedules. "We're allowed two quick looks out the window," Gene Cernan reminded Harrison Schmitt as they prepared to descend to the moon's surface during Apollo 17.

Lee stops the Humvee and consults the GPS. We've reached our first "way point." It's a geology pit stop: don spacesuits, climb a bluff, collect samples. Lee and Abercromby are standing outside the vehicle, fiddling with their communications headsets, which enable them to speak to each other and to "ground control," back at the HMP base. Around the rear of the Humvee, Chase has laid out simulated suit components on two mats. If this were the actual rover, the suits would be hanging off a pair of suit ports cut into the vehicle's rear panels. The astronauts would step into them from inside the rover, twist their torsos to unlock suit from port, and walk away. And then reverse the process when they return, leaving their suits dangling like shed exoskeletons. This way the suits don't clutter the cramped interior, and no dust gets inside.

Dust is the lunar astronaut's nemesis. With no water or wind to smooth them, the tiny, hard moon rock particles remain sharp. They scratched faceplates and camera lenses during Apollo, destroyed bearings, clogged equipment joints. Dusting on the moon is a fool's errand. Unlike on the Earth, where the planet's magnetic field wards off charged particles of solar wind, these particles bombard the moon's surface and impart an electrostatic charge. Moon dust clings like dryer socks. Astronauts who stepped from the Lunar Module in gleaming white marshmallow suits returned a few hours later looking like miners. The Apollo 12 suits and long johns became so filthy that at one point, astronaut Jim Lovell told me, the crew "took off all their underwear and they were naked for half the way home."

Another reason to keep moon dust outside the rover: With so little gravity, inhaled particles may settle more slowly and thus penetrate farther into the lungs, reaching the more vulnerable tissue deeper in. NASA has been funding so much research on dust and dust mitigation that an entire lunar dust simulant industry exists.* (Moon rocks and pebbles are classified as "national trea-

---

* NASA buys it by the ton, but you can buy it by the kilogram ($28). Go to the eNasco educational products Web site, but not if you're squeamish. "Save on Lab Time!" says the promo copy for skinned cats. The eNasco dissection specimen section offers ten different skinned cat products, proving that there is, in fact, more than one way.

sure" and can't be sold, but no such prohibition applies to moon dust, real or simulated. Which explains why a dust-coated Apollo 15 mission patch sold for $300,000 at a 1999 Christie's auction.)

Lee considered cutting holes in the back of the Humvee and trying to rig a pair of mimic suit ports for this week's simulations. Weaver was aghast. "I told him, 'You are *not* cuttin' up the Humvee.'" The HMP mechanic is a high school student from Tennessee, barely shaving but possessed of a scraggy, hard-shelled sang-froid. Lee, who knows Weaver's mother, saw him rebuilding a dirt bike motor and offered him the greatest summer job in the history of summer jobs.

Lee genuflects on one of the mats while Chase prepares to lower the simulated PLSS (portable life support system—that bulky white astronaut backpack) onto Lee's torso. His arms are outstretched, as though in supplication, or delivery of a Broadway musical number. Chase's employer, Hamilton Sundstrand, makes both real and simulated spacesuits, both of which require valets. (One of the less heroic aspects of a spacewalk: Someone will have to help you pull up your pants.)* As Chase and Lee grapple with the PLSS simulator, Weaver takes a pack of Camels from a pocket.

---

* And you will wear a diaper. These days it's called a maximum absorbent garment. The MAG replaces the DACT (disposable absorbent containment trunk), which had less (not enough) capacity. In the Apollo era, astronauts wore both a pull-up fecal containment device (FCD) and a condom-attached urine containment device. Let's let astronaut Charlie Duke, providing commentary for NASA's Apollo 16 Lunar Surface Journal, explain the system: "[The FCD] was like a ladies' girdle you pulled on and it cut out in the front so that your penis could hang out so you could get on the UCD . . . I think there was maybe a jockstrap that went on, also, and it had a hole for your penis, and then you rolled on your UCD and then you buttoned that or snapped it to the jockstrap."

EVAs, to him, are more or less cigarette breaks. He's leaning toward a career in flight, but as a bush pilot, not an astronaut.

Since there's oxygen in Canada, you may be wondering what's inside a simulated life support backpack. A fan, mostly, to keep the helmet faceplate from fogging. It doesn't much matter what's in it. The idea is to burden the wearers and restrict their movements and field of view in the same manner that an astronaut will be burdened and restricted. Then give them some tools and tasks and see what sorts of problems develop.

As on Apollo, tasks are written on a pad Velcroed to the cuff of the spacesuit. Outer space is list world: cuff checklists, lunar surface checklists, lists of mission rules and "get-ahead tasks." Morning in orbit begins with a fax or email of the day's schedule and tasks, updated with last-minute changes. Any deviation must be reported to Mission Control. Outside of the hour or two designated as "pre-sleep," every waking hour is planned. It's like a book tour.

Abercromby is flipping through his cuff checklist. He has laminated it, because it rains a lot on Devon Island and because he has a head for planning. I don't know much about Abercromby, or NASA for that matter, but from what I've seen, I could imagine him running the place one day. He is taking these simulations very seriously. His 66-page Field Test Plan includes time lines, objectives, a four-page hazard analysis, an Off-Nominal Situation Resolution Tree and, for each simulated traverse, science priorities, targets of opportunity, get-ahead tasks, and mission rules. The document has been distributed to, but possibly not read by, everyone participating.

Abercromby steps into a set of the white Tyvek coveralls that are standing in for pressure suits. Ping Pong is biting one of Lee's gloves and dancing around the men's feet. "Does Ping Pong want to go EVA?" Lee is using his special, high-pitched Ping Pong voice. Abercromby interrupts them. "We should talk about get-ahead tasks and targets of opportunity."

Weaver watches through smoke. "You look like a crew of painters."

Once the helmets and life support simulators are on, Chase shoots some video. Abercromby looks mildly uncomfortable. Lee has no problem with the getup. Even a pretend spacesuit, I'm told but have some trouble believing, is a chick magnet. Lee, forty-five, is single and something of a heartthrob in the space community.

Rock hammer in hand, Lee heads up the slope of a hill. Abercromby follows with a sample bag. The teams' tasks are modeled on Apollo-era EVAs—selecting and bagging rock and soil samples, photographing, and taking gravity meter and radiation readings.

Only one Apollo astronaut, Harrison Schmitt, was a geologist. The rest were pilots who had been given a crash course in lunar geology to help them know what to look for and how to read the land. The training included time in a NASA geology lab with Earth basalts and breccias, painted Styrofoam moon rock mock-ups, and, after Apollo 11, actual lunar samples. Field trips took them out to the Nevada Test Site, 65 miles northwest of Las Vegas, where the Atomic Energy Commission tested nuclear bombs in the fifties, leaving craters up and down the desert floor. Because the rocks were still radioactive, the astronauts couldn't pick them up and examine them. No one seemed to care, as they were, recalls Jim Irwin in the astronaut commentary of the Apollo 15 Lunar Surface Journal, "anxious to get back into Las Vegas."

One focus of today's traverse is timing. How closely are the rovers able to stick to the time line? How often should they check in with ground control, and how do you update the plan on the fly, if one group falls behind? The teams have been asked to keep track of the start and stop times for each phase of the traverse, to see whether things are taking longer than predicted, and if so, what's slowing them up. At some point, intern Jonathan Nelson will deliver a "productivity metrics" report that will make some NASA manager feel calmer about the $200,000 budget he or she authorized for Arctic analog projects this summer. For now, it means lots of conversations like this one:

NELSON: What do you want, suit time?
LEE: No. Basically, when we started to suit up . . .
NELSON: So you want suit time.
LEE: That's what suit time is?
NELSON: There's a difference between prep and suit.
ABERCROMBY: So what was our boots-on-the-ground time?

Timing is critical to an astronaut wandering around on an extraterrestrial surface. Without knowing how long it takes to walk or drive a given distance on a certain kind of terrain, it's hard to know how much oxygen or battery life one will need. Apollo astronauts had to conform to "walkback constraints." These were, and are, first figured out by driving someone out on some lunar analog terrain, say, 3 miles from base, putting a suit simulator on him, marking the start time, and letting him walk back. Apollo astronauts were not allowed to drive farther from the safety of the Lunar Module than the distance they could walk without running out of oxygen, in case the rover broke down. (This is a rationale for having two rovers; if one malfunctions, the other can come pick up the stranded crew.)

Walkback constraints were a source of worry for Apollo mission planners and frustration for astronauts. Without trees or buildings to give a sense of scale, it was difficult to accurately estimate distances. For safety's sake, estimates were conservative, sometimes maddeningly so. On the way back from an Apollo 15 EVA, astronaut Dave Scott spied an unusual black rock sitting out on its own. He knew that if he asked Mission Control for permission to go get it, they'd tell him to keep driving, as the EVA was already behind schedule. Since Mission Control could hear their conversations, Scott fabricated a seatbelt malfunction. The rock would become known as "the seatbelt basalt."

SCOTT: Oh, there's some vesicular basalt right there, boy. Oh, Man! Hey, how about ... Let's just hold on one second, we've got to have . . .
IRWIN: Okay, we're stopping.
SCOTT: Let me get my seatbelt. . . . It keeps coming off.
IRWIN [picking up on the ruse right away]: Why don't you hand me your seatbelt?
SCOTT: Just a minute . . . If I can find it. [pause] There it is. [pause] If you'll hang on to it here for a second.
IRWIN: Okay, I've got it. [long pause]

It's late afternoon now. We've reached the end-of-the-day rendezvous point. Lee and Abercromby will overnight here, on primitive bunks in the back of the Humvee, while the rest of the team drives back to camp and then rejoins them in the morning. Bravo Party is nowhere in sight, so we wander over and take pictures of each other standing on the lip of a ravine. Later, I'll look at these photographs and it will appear that I was visiting a strip mine. It's hard to say why I find Devon Island beautiful. But there are these moments when you're tromping along, head lowered against the wind, and your eye lands on a hump of moss with tiny red flowers like cupcake sprinkles, and you're just walloped by the sight. Maybe it's the unlikely heroics of something so delicate surviving in a place so stingy and hard. Maybe it's just the surprise of color. At one point yesterday, on a hike through yet another grey and beige canyon, a bumblebee flew past. The yellow seemed like a hallucination, something colorized in a black and white photograph. "Whoa, buddy," someone said. "Where'd you take a wrong turn?"

It's starting to rain, so we head back to the Humvee. Lee and Abercromby are in high spirits, having completed day one of NASA's very first pressurized roverlike traverse. "Just terrific," Abercromby is saying. "There can't be many places in the world where the terrain and the scale so closely approximate lunar—"

"Ground, this is Bravo Party." It's the radio. NASA geophysicist Brian Glass, the SPR-Bravo traverse leader, reads out his GPS coordinates and a weather update. *Read* is the wrong verb. It's something between *shout* and *spit*. It's raining hard where they are. Their visibility is down to 300 feet. Bravo Party isn't in a Humvee. Their rover simulator is a Kawasaki Mule, a larger ATV with a short pickup bed. Their spark plugs got wet crossing streams that had appeared shallower in satellite photographs. One of the spare plugs was the wrong size. At one point, they were almost two hours behind.

Weaver flips his hood over his head. "Sounds like the other guys aren't havin' as much joy."

MORNING AT HMP begins with the sound of tent zippers. The sleeping accommodations are thirty nylon tents, hunkered on a hill, breaking rank with the island's color scheme. Everyone gets up around the same time, because every morning begins with a meeting. This morning's is being held in the main office tent. Along with the NASA meeting mentality, an actual NASA phone system has been set up on Devon Island. Staff at NASA Ames, in California, can dial a four-digit extension and reach Lee, a couple hundred miles from the magnetic North Pole, on an in-house call.

(HMP is one of those odd but surprisingly common Internet-age locales with VoIP coverage but no flush toilets.)*

An HMP webcam is set up on a tripod in one corner, enabling people all over the world to look on as Andrew Abercromby attempts to maintain order and civility at the posttraverse Lessons Learned Review. One of HMP's ancillary research goals is the study of "human dynamics which result from extended contact in close quarters." Hopefully someone other than myself is taking notes this morning.

"No one told us we were behind after the first EVA," Glass is complaining. "According to the paper time line we were ten minutes early." Something about Glass's receding red hair and the cut of his mustache and beard make me think of Sir Walter Raleigh. It's easy to picture him with an Elizabethan collar atop the polar fleece. Glass says ground control made them wait nearly two hours while they mapped a quicker route. "I . . ." He exhales. "I had the impression we were being jerked around just so Alpha Party could get back in time for dinner."

Lee insists that Alpha Party had had no idea any of this was happening.

"Well, yes," Glass says, "because . . ." He turns to Abercromby. "Pascal had his iridium phone set to Ignore."

"It was on Vibrate!"

---

* In this case, to keep the island more Mars/moon–like. (Biowaste encourages plant growth.) Fourteen 50-gallon drums of urine are flown off the island each season. Men go directly into the drum via a funnel. Women squat over a pitcher first. It's one of those clear plastic pitchers they use for beer at campus pubs. Pouring it out was like an entire Saturday night of drinking condensed in a single gesture. Solid waste happens on a toilet seat mounted over a plastic bag that you then take away and drop in the trash. You are your own dog.

"Can we," says Abercromby, "try to drive toward lessons learned?"

Glass has moved on to "the seemingly incessant" calls from ground control to check in on what they were doing. "Every time, I had to stop, get to a place with no wind noise and no motor noise, take off the helmet . . ."

Lesson learned: Explorers appreciate a little autonomy. The rigidly scheduled time lines that typify shorter planetary surface EVAs will have to loosen if NASA pushes ahead to two-week EVAs and trips to Mars. Autonomy is the topic of the moment among space psychologists. Astronauts often complain to flight surgeons about not being allowed to make their own schedules and decisions about their work. Like Glass, some find Mission Control's micromanagement frustrating and demoralizing. Space psychiatrist Nick Kanas, of the University of California, San Francisco, has studied the psychological effects of high and low autonomy on personnel in three different space simulations. The men and women Kanas studied were generally happier and more creative in the high-autonomy scenario. The exception was the guys in Mission Control, who "reported some confusion about their work role."

The meeting shows no sign of abating. Weaver is in presleep mode. The HMP field guide, known for his laissez-faire shower regimen, is scratching his back on the doorframe like a molting grizzly. Glass isn't quite done. ". . . We had no lunch other than candy bars. Alpha Party had taken multiple items that—"

"No way," says Lee. "We had a total of two sandwiches."

"Lessons learned," Abercromby says flatly. "Order more bread."

Mike the cook speaks up. "Some bread got stolen in Resolute." (Flights to Devon Island leave from the Inuit hamlet of Resolute.) Mike had three days to singlehandedly plan meals and buy supplies for thirty-some people over a six-week field season.

The NASA traverse planning office should probably hire Mike the cook. One of the problems with expedition planning today, versus forty years ago, is that NASA is so much larger. Too many cooks take forever to agree on how to make the broth. Or as Apollo mastermind Wernher von Braun is said to have commented on the moon landing, "If we'd been more people, we'd have failed."

Gene Cernan, in the astronaut commentary for the Apollo 17 Lunar Surface Journal, bemoans the endless prepping and what-iffing that typifies today's NASA. "I don't know if we . . . have the mentality—I don't want to say 'guts'—to take the kind of risks we did when we [went to the moon] the first time. . . . And that's a sad commentary." After all, no matter how much you plan and how carefully you engineer things, there will always be problems. The safety manager of the eighth Apollo mission once famously pointed out: "Apollo 8 has 5,600,000 parts. . . . Even if all functioned with 99.9 percent reliability, we could expect 5,600 defects."

On the other hand, as they say, failing to plan is planning to fail.

Years ago, I interviewed astronaut Chris Hadfield for an article about how crews train for spacewalks (EVAs wherein astronauts float outside the spacecraft, usually to make repairs or add new hardware). I asked him if he thought NASA overdid it with their protracted rehearsing and planning. Hadfield would spend 250 hours in the Neutral Buoyancy Laboratory practicing for a six-hour EVA. (The NBL is a huge indoor pool containing ISS mock-up pieces; floating in a spacesuit in water is a passable approximation of spacewalking.) "Yeah, there's lots of options," Hadfield said. "You could do nothing and hope for the best, or you could spend billions of dollars on each flight and try to nail down every last detail." NASA, he says, aims for somewhere in the middle. "The prep is what matters," he added. "That's what we do for a living. We don't fly in space for a living. We have meetings,

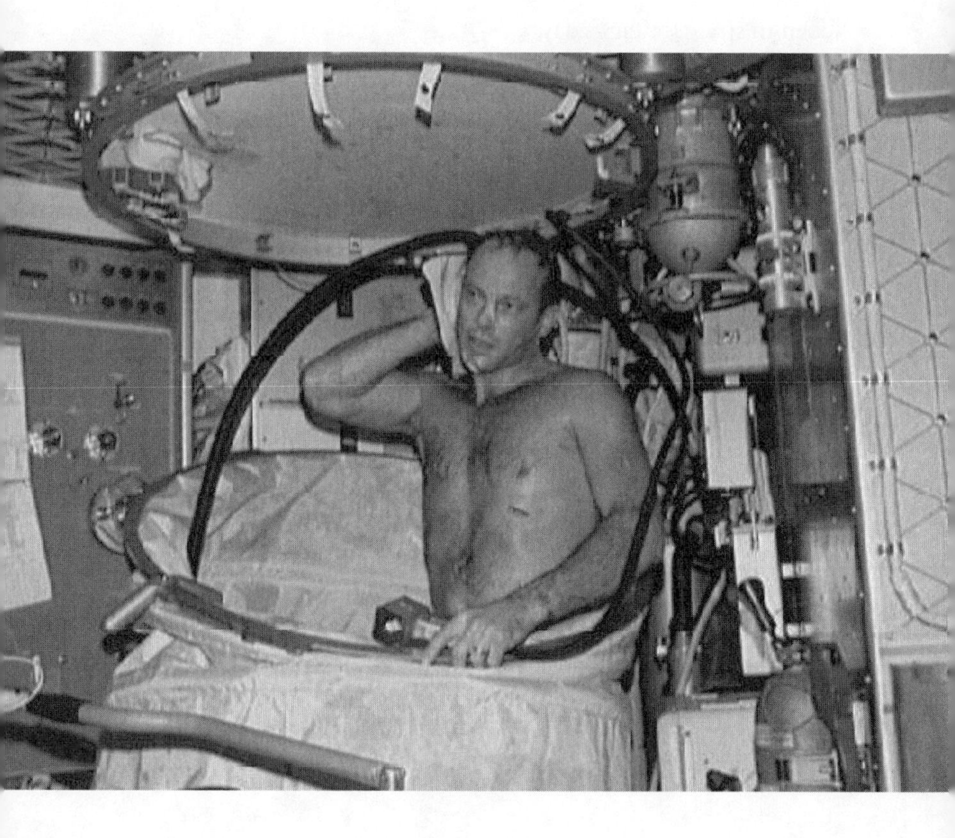

10

HOUSTON, WE HAVE A FUNGUS

Space Hygiene and the Men Who Stopped Bathing for Science

Jim Lovell is best known as the commander of Apollo 13, the astronaut with the problem. As anyone who's seen the Tom Hanks movie knows, an oxygen tank exploded on the way to the moon, knocking out power in the Command Module and forcing Lovell and his two crewmates to hunker down in the Lunar Module for four days with limited oxygen, water, and heat. For forty years, people have been coming up to Lovell saying, "My god, what an ordeal." I said that to him too, but not in reference to Apollo 13. I was talking about Gemini VII: two men, two weeks, no bathing, same underwear. Inside a pressure suit, inside a capsule so cramped that Lovell could not straighten his legs.

Gemini VII, launched on December 4, 1965, was a medical dress rehearsal for the Apollo lunar program. A round-trip moon mission takes two weeks, and no astronaut had spent that much time in zero gravity. (NASA's record at that point was eight days.) If a medical emergency was going to develop on, say, the thirteenth day, the flight surgeons wanted to learn about it when the astronauts were 200 miles from Earth, not 200,000.

There was concern that wearing a spacesuit for two weeks in a space the size of the front seat of a VW Beetle might be unendurable. The ever-heedful NASA proposed to Lovell and his crewmate Frank Borman that they undertake a real-time simulation of Gemini VII inside a mock-up of the capsule—a rehearsal rehearsal. "Fourteen days sitting in a straight-up ejection seat on Earth?" says Borman in his NASA oral history. "We were able to get that nonsense kicked out in a hurry."*

In fact, there was no need for the nonsense, because similar nonsense was already underway out at Wright-Patterson Air Force Base in Ohio. From January 1964 to November 1965, a series of nine experiments on "minimal personal hygiene"—including a two-week Gemini VII simulation—had been taking place in an aluminum space capsule simulator inside Building 824 of the Aerospace Medical Research Laboratories. The AMRL people did not mess around. *Minimal* was defined as "no bathing or sponging of the body, no shaving, no hair and nail grooming . . . , no changing clothes and bed linen, the use of substandard oral hygiene, and minimal use of wipes" for, depending on the experiment, anywhere from two to six weeks. One team of subjects lived and slept in spacesuits and helmets for four weeks. Their underclothes and socks deteriorated so completely that they had to be replaced. "Subject C became so nauseated by body odor that he was forced to remove his helmet after wearing it for less than ten hours. Subjects A and B had already removed their helmets by that time." It didn't help. With the helmet off, body odors were "forced

---

* Borman could be a bit crusty. As Lovell put it, "Two weeks with Frank Borman anyplace is a trial."

out around the neck of the pressure suit," a situation described by B, on day four, as "absolutely horrible." This explains why Frank Borman, in the mission transcript for the second day of Gemini VII, asks Lovell if he has a clothespin. He's about to unzip his suit. ("For your nose," he tells the perplexed Lovell.)

For a different set of subjects, the heat was turned up to 92 degrees Fahrenheit. The Gemini VII simulated crew not only spent two weeks, day and night, in a spacesuit, but had to struggle with the same waste collection systems that would soon bedevil Lovell and Borman.

To quantify the squalor, the Air Force scientists would usher the men—most of them students from the nearby University of Dayton—into a portable shower, one by one, and collect the runoff for analysis. John Brown was the officer in charge of the simulated space capsule, formally known as the Life Support Systems Evaluator and casually known as "the chamber." Oddly, the showers were the part Brown recalls the men complaining about. Because the water was unheated. "They didn't want the hot water cooking the skin flakes," he said, speaking four words together that have no business being so.

As unsavory as this project must have been for the subjects, it was no bowl of rose petals for the researchers. It was their meandering sniffs that made possible the conclusion: "Body odor strongest in axilla, groin, feet."

Axilla (armpits) and groin occupy the top two slots because that's where the body's apocrine sweat glands are. Unlike the body-cooling eccrine sweat glands, which secrete mainly water, the apocrine glands produce a cloudy, viscous secretion that, when broken down by bacteria, creates the hallmark BO punch. I don't know quite how to phrase this or what it reveals about me, but I have never detected BO in the pubic region. O, for sure, but not BO. I asked University of Pennsylvania dermatologist and body odor researcher Jim Leyden about this. He verified the apocrine presence in the groin, and insisted there's a similar smell. "It's just not that easily appreciated," he said, "because the sensing device is farther away." I decided to let it ride.

The apocrine glands are hooked up to the autonomic nervous system; fear, anger, and nervousness prompt an upswing in secretions. (Companies that test deodorants call it "emotional sweat," to distinguish it from the temperature-triggered kind.)* You would think that being strapped to a launching rocket would be a situation in which a man would be, to quote Leyden, "milking those glands for everything they're worth." I asked Jim Lovell, in a telephone conversation, if he could recall the comments made by the frogmen who opened the Gemini VII hatch after splashdown.

"You're investigating a rather unusual aspect of spaceflight," he said. He didn't remember, but he did recall comments made by some of the Apollo hatch-openers. "They'd get a whiff of the inside of that spacecraft and it smelled . . ."—Lovell's gentlemanly instincts intervened—"different than the fresh ocean breezes outside."

Underarm sweat supplies both food and lodging for bacteria. Eccrine sweat is mostly water; it provides the moisture bacteria need to thrive. Protein-rich apocrine secretions are the twenty-four-hour diner. (Though eccrine sweat does contribute edible elements whose breakdown products are, as Leyden says, "part of the overall bouquet, if you will." It's a milder, lockerroomy smell.)

The armpit is not entirely the bacterial paradise it would seem to be. Sweat has natural antimicrobial properties. Though they

---

* That is why some deodorant and antiperspirant efficacy tests include an "emotional collection." A group of subjects sit with pads under their arms to absorb secretions while being forced to sing karaoke or speak in front of a group. The pads are then weighed and the armpit smells rated by professional odor judges. I was once, as part of an article on body odor, invited to be a guest judge. "Take little bunny sniffs," I was told.

don't by any means render the skin sterile, there are limitations to what can grow there. That may be one reason why the Air Force boys' odors hit a plateau, rather than growing ever worse as the weeks wore on. The technical report states that the men's body odor reached its "maximum height" at seven to ten days, and then began to subside. Height is an odd attribute for smell, but it's possible to imagine how in this case the odor could seem to be taking on physical proportions, growing taller, sprouting heads, limbs, quills.

Soviet space biologist V. N. Chernigovsky, in 1969, carried out a restricted-bathing experiment of his own, this one including bacteria colony counts. The bacteria populations in subjects' armpits and groins plateaued somewhere between the second and third weeks. At which point there were roughly three times as many colonies as on freshly washed skin. (Except on the feet* and buttocks, where there were seven to twelve times as many.) A Navy study turned up similar findings; here some subjects' bacteria counts even began to drop after two weeks.

The other explanation for the odor plateau is that the men's body odor had become so strong that it was impossible for whoever was judging it to detect incremental changes. Weber's Law provides the explanation. The detection threshold for changes in a particular smell (or sound or sensation) varies according to the intensity of the background smell (or sound or sensation). Say you are in a noisy restaurant. If the noise level rises a few decibels, you can't tell. Had the room been quiet, you could easily tell. If someone's armpits have been shouting for a few days, it's hard

---

* Because of all the sweat and dead skin (calluses), the bottoms of the feet and the spaces between the toes are a Mecca for bacteria—high numbers, much more variety. One class of dead-skin-eating bacteria, *L. brevis*, excretes compounds that smell like ripe cheese. Though it may be technically more accurate to say that certain ripe cheeses smell like feet: Cheesemakers routinely inoculate certain of their creations with *L. brevis*.

to tell when they're shouting a little louder. Jim Leyden gives the example of his son, who was a rower in college. One year the team decided they were going to wear the same rowing outfits until they lost. "Well, they became national champions that year. You could not get near that boat. The smell may have plateaued, but as far as I was concerned, it was just constantly horrible."

Eventually the mind stops registering the body's smell. In Leyden's words, "It's going, 'I don't need to bother telling you this anymore.' " Unfortunately for a group of AMRL subjects in a twenty-day no-bathing Apollo simulation, this point didn't arrive until day eight.

NASA would have done well to add body odor anosmia to its list of desired astronaut traits. Some people* are genetically unable to smell (i.e., they're anosmic to) one or both of the two BO heavies: 3-methyl-2-hexanoic acid and androstenone. "Have you ever been on an elevator with someone and wondered, 'How can he come on here smelling like that?' Well, he may be anosmic to his odor," Leyden says. "And those of you who have never experienced that, you may be one of those people on the elevator that everyone's wondering about."

Aside from body odor, the most common contributor to what one researcher called "perceptions of personal dirtiness" is not dirt per se, but bodily emanations that have built up on the skin: grease, sweat, and scurf,† to be specific. Where you have hair, you

---

* And possibly deer. A 1994 issue of *Crop Protection* details the failed but entertaining efforts of botanists at the University of Pennsylvania to deter white-tailed deer by dousing an assortment of ornamental shrubbery with 3-methyl-2-hexanoic acid. Which raises the unusual marketing question, Will a homeowner abide a rhododendron that smells like BO?

† A.k.a., shed skin. *Dorland's Medical Dictionary* defines scurf as "a branny substance of epidermic origin"—an evocative pairing of dander and breakfast cereal. *Try new Kellogg's Dandruff Flakes!*

have sebaceous glands; that is to say, everywhere but your palms and the soles of your feet, where greasiness is a slip, trip, and fall hazard and thus a survival liability.

The 1969 Soviet restricted-hygiene experiments monitored the build-up of oils, or sebum, in male volunteers. (Here, in addition to not bathing, the subjects had to spend "most of their time sitting in an armchair." The simulated astronaut of the sixties was a stinky guy watching TV in a dirty undershirt.) For the first week without bathing, the skin's oiliness remained constant. Why didn't it increase? Because clothes are surprisingly effective absorbers of sebum and sweat. The Soviet researchers collected wash water from the subjects' skin in one basin, and wash water from their clothes in another. They compared the amounts of grease, sweat, and dander in the two tubs. Eighty-six to 93 percent of the skin's emanations were in the water where the clothing had been. In other words, all but 7 to 14 percent of the men's filth had been absorbed by the fabric of their clothing. This was true of cotton, cotton-rayon blend, and, to a lesser extent, wool.

The Soviet findings help explain the lackadaisical hygiene practices of the sixteenth and seventeenth centuries. Renaissance doctors discouraged washing with water. Removing the protective layer of oil from the skin, they believed, left the bather vulnerable to plagues, tuberculosis, and a host of other ills then believed to spread via "miasmas" that seeped into the body through the pores. Queen Elizabeth I, her era's version of a clean freak, famously wrote, "I bathe once a month, whether I need it or not." Many let it go a year.

But here's the thing: Instead of showering once or twice a day, Renaissance men and women would change their undersmocks and chemises. The men of Gemini VII and the AMRL chamber, on the other hand, couldn't change their underthings. The authors of the AMRL chamber study noted that the subjects' clothes eventually began "sticking to the . . . groin and other body fold areas and were very odorous and starting to decompose," a condition described as "very troublesome." Lovell told me the Gemini VII long johns were in bad shape by the end of the mission. "They were," he allowed, "pretty smudged around the crotch area"—even more so than those of the average person who didn't bathe or change his underwear for two weeks, because the average person wasn't testing a new NASA urine management system that "leaked considerably sometimes." For instance, on the second day of the flight, when Lovell, reporting to Mission Control that he was ejecting urine from the spacecraft, noted, "Not too much of it; most of it's in my underwear."

At a certain point, clothes reach their saturation point, and sebum begins to accumulate on the skin. According to the Soviet researchers, who monitored oil levels on subjects' chests and backs, it takes five to seven days for a cotton garment to reach this point. It is difficult to pinpoint the day when the Gemini VII astronauts began to notice the buildup on their skin. By day ten, they were "starting to itch" and "getting a little crummy" in the scalp and crotch. Here they are on day twelve:

MISSION CONTROL: Gemini VII, this is Surgeon. Frank, do you have any lotion remaining?

BORMAN: Any lotion?

MISSION CONTROL: Roger.

BORMAN: We have some but we sure don't need it, Jack. We are as greasy as can be.

It is unusual to come across the word *lotion* in a NASA mission transcript. Borman seemed nettled by NASA's preoccupation with skin care, as though it were compromising the overall manliness of the mission. At one point, the flight surgeon comes on the microphone to ask, "And how are your skins?" Earlier, he'd accosted Borman out of the blue with the inquiry, "Are you having any difficulty with drying of your lips?" "Say again please?" answers Borman. You get the sense he heard him fine. On the fourth day, Mission Control fixated on how much Borman was perspiring. Borman, like his epidermis, had reached the saturation point. He refused to answer, forcing Mission Control to try to enlist Lovell's help.

MISSION CONTROL: Do you notice in looking at him that his skin is moist?
LOVELL: I'll let him answer that.
BORMAN: [silence]
MISSION CONTROL: Have you been sweating at all, Frank?
BORMAN: [silence]
MISSION CONTROL: Gemini VII, this is Carnarvon. Did you copy?
BORMAN: About sweating? I'd say, yes, I'm perspiring a little.
MISSION CONTROL: Very well. Thank you.

Once a set of clothes becomes saturated and oil starts to build up on the skin, what's the end point? Does uncleansed skin grow ever greasier as the days pass? It does not. According to the Soviet research, the skin halts its production of sebum* after five to seven days of not bathing and not changing one's increasingly well-greased clothing. Only when the person changes his shirt or takes a shower do the sebaceous glands get back to work. Skin seems

---

* Roughly 4.2 milliliters per day, according to a table in a paper by Mattoni and Sullivan, entitled "Synopsis of Weight and Volume of Waste Product Generation from All Sources in the Closed Environment of a High Performance Manned Space Vehicle." That is just under a teaspoon of skin oil, an equivalency made with the help of a recipe conversion table. Employed in tandem, the two tables would enable the deranged or geographically isolated baker to substitute sebum for vegetable shortening or calculate the equivalent of a cup of flour in desquamated epithelium.

happiest with a five-day buildup of oils. Listen to Professor Elaine Larson, editor of the *American Journal of Infection Control,* talking about the stratum corneum, the outermost layer of human skin: "This horny layer has been compared to a wall of bricks (corneocytes) and mortar (lipids)" and helps "maintain the hydration, pliability, and barrier effectiveness of the skin."

Do we compromise our skin's health by constantly scrubbing off the mortar? Does our skin want us to bathe every five days? Hard to say. It's true that especially zealous hand washers—hospital personnel and certain obsessive-compulsives—often develop irritation and eczema. Twenty-five percent of the nurses in one study, writes Larson, had dry, damaged skin. Ironically, the nurses may be exacerbating the very thing that hand-washing seeks to prevent: the spread of infectious bacteria. Larson says healthy skin sheds 10 million particles a day, and 10 percent of those harbor bacteria. Dry, damaged skin flakes off more readily than healthy, lubricated skin and thus disperses more bacteria. Damaged skin also harbors more pathogens than healthy skin. As Larson says, "Perhaps sometimes clean is too clean." Most Americans don't wash often enough to cause skin problems, but they certainly wash more than necessary. In the words of some academic I can't name because I've lost the first page of his paper, "Personal hygiene as practiced in the U.S. today is largely a cultural fetish, actively promoted by those with commercial interests."

In space, as in the military, bathing is more an issue of morale than of health. Space agencies, recognizing what one researcher called "the psychological inadequacy of sponge baths," devoted a lot of time and money in the 1960s trying to develop a zero-gravity shower for space stations. One of the earliest prototypes tested was a "shower suit." The technical report I read included the following less-than-encouraging summary: "Results leave much to be desired in the showering, rinse, and drying procedures." The usual arrangement doesn't work; the water sprays from the shower head for a few inches and then collects in an expanding blob: fascinating, but of little ablutionary help. If you hold the shower head close enough to forestall the big blob, then the water ricochets off your skin, forming floating drops that you then have to spend ten minutes chasing down to keep them from floating out into the station. "It turned out to be easier just to forget the whole thing," said astronaut Alan Bean, of the collapsible Skylab shower.

The shower on the Soviet space station Salyut used air flow to try to draw the water down toward the cosmonauts' feet. It was minimally successful. Blobs formed, and blobs tend to cling to the body's concavities, including the mouth and nostrils. To keep from choking, cosmonaut Valentin Lebedev and his crewmate Tolia Berezovoy wore snorkeling gear. "What an exotic sight it was," wrote Lebedev in his diary. "A naked man [flying] across the station, . . . with snorkel in his mouth, goggles over his eyes, a clip on his nose." Understandably, the crew of Salyut 7, like Elizabeth I, showered just once a month. These days there are no space showers. Astronauts wipe themselves with moistened towels and rinseless shampoo.

Bathing is more important on the space stations, because the missions are longer and they include daily exercise regimens that ratchet up the sweat level. As an adjunct to body wiping, Japanese astronauts on the ISS have been wearing "J-Wear," developed at Women's University in Tokyo out of fabric "with the function of dissolving foulness and body odor by photocatalyst and prevention of the rotten smell of sweat by the antibacterial nano-matrix finishing technique." Astronaut Wakata Koichi (pronounced, perhaps aptly, *co-itchy*) wore the same J-Wear underpants for twenty-eight days without complaint.

The astronauts of Gemini VII could only dream of "comfortable everyday clothes for life in a spaceship," as one press release calls J-Wear. They wore hot, heavy, bulky spacesuits for pajamas.

The subjects in the Air Force Gemini VII simulation were plagued by "chafing and much irritation in groin." In case you have ever questioned the value of thorough wiping and regular changes of underwear, here's a reason. In people with poor bathroom habits or 1960s Air Force hygiene restrictions, fecal bacteria migrate. Wright-Patterson researchers sampled thirteen sites on the men's bodies to check for *E. coli*. It was a remarkable Diaspora. Fecal bacteria had made its way to the men's eyes, ears, and, in two cases, toes. Five out of six of the Soviet subjects who sat in armchairs for thirty days developed folliculitis—bacterial infection in the hair follicles on the skin. Three developed boils—especially bad, swollen, painful, infected follicles. (The Soviet paper uses the old-timey term "furuncle." You almost want one just to be able to go around saying "furuncle.")

Lovell doesn't recall any skin problems. "The difference is zero G," he told me. "That's the key to the whole deal." When a man floats a few inches above his chair, when his arms hover out away from his sides, he has less of the chafing and irritation normally caused by damp, filthy clothes rubbing sweaty, unwashed skin. The astronauts' underwear didn't get plastered to their buttocks. Whatever bacteria lurked in their sweat, it wasn't getting ground into their follicles. There is a condition called hot-tub folliculitis, which often appears on hot tubbers' buttocks and the backs of their thighs—just where the friction and pressure is. (The water in a hot tub is hot, but not hot enough to kill bacteria. An undertreated hot tub is essentially, quoting University of Arizona microbiologist Chuck Gerba, " *E. coli* soup.")

DAY SIX OF GEMINI VII. Frank Borman is on the mic. The exchange is proceeding in the macho, jargony manner of pilot-to-ground communications. Until:

MISSION CONTROL: Stand by for the Surgeon, Gemini VII.

BORMAN: [silence]

MISSION CONTROL: Gemini VII, this is Surgeon. Have you had any dandruff problem up there, Frank?

BORMAN: No.

MISSION CONTROL: Say again.

BORMAN: N. O. No, negative!

Commander Borman did not wish to discuss skin care. But later, in his memoir, he would write about "our scalps" and about the case of "terminal dandruff" he had. Though it probably wasn't, technically speaking, dandruff. Dandruff is caused by an inflammatory skin response to oleic acid, which the scalp fungus *Malassezia globosa* excretes after dining on your scalp oils. Either you're sensitive to oleic acid or you're not. If Borman didn't have dandruff before he went into space, he didn't have it afterward, says dermatologist Jim Leyden. Leyden once paid prisoners to not wash their hair for a month, specifically to see if they developed dandruff. They did not. The flakes on Borman's head and skin were most likely the accumulation of millions of shed skin particles—particles normally washed away in the shower—mixing with sebum and clumping together.

The atmosphere in Antarctic field camps is similarly dry and shower facilities similarly nonexistent or cumbersome, making the six-week Antarctic Search for Meteorites field season a good analog for space hygiene. "Six weeks of dead skin is like two whole layers," says team leader Ralph Harvey. Sometimes it all comes off at once, in the first wash. Harvey admits to being fascinated by the spectacle. "I remember coming back and taking a shower and the whole end cap of my finger would just come off."

What makes the dander situation bearable in Antarctica is that you can step outside your domicile and shake out your long johns and sleeping bag. You can't do this in space or simulated space. The description of the Navy space cabin simulator at the end of the experiment was like a ski report. "A fine layer of powdery scales was found to cover the floor of the chamber."

In zero gravity, the flakes never fall. I asked Lovell about this. I believe my exact words were, "Was it just like a snow globe in there?" He said he didn't recall anything like that. Or not "of such magnitude that it would stick in my mind all these years." (For the thing that did stick in his mind all these years, see chapter 14.)

The head in general is a problem. The majority of our sebaceous glands are attached to hair follicles, thus the unwashed scalp quickly becomes a greasy thing. So much so that the bathphobic hordes of the sixteenth century would rub powder or bran into their scalps before retiring for the night, much as homeowners today sprinkle kitty litter on motor oil spills. Like sweat, sebum develops a distinctive aroma as bacteria break it down. "At least two of the Skylab astronauts reported that their heads developed offensive odors," noted space psychologist Jack Stuster in a 1986 NASA report on space station habitability.

BORMAN AND LOVELL did not stay in their suits the entire flight, as NASA had originally planned. On day two, flight surgeon Charles Berry began lobbying NASA management on their behalf. A compromise was struck: Only one man had to stay suited (in case of a depressurization emergency). Borman drew the short straw, and Lovell squirmed out of his suit. For years, Lovell recalls, his son would tell friends, "Dad orbited the Earth in his underwear!"

By hour 55, Borman has his suit unzipped and halfway off. By hour 100, he petitions NASA management to let him take it all the way off. Five hours pass. Houston comes back on the line. Borman may take off his suit, but only if Lovell gets back in his.

Lovell tries to resist ("I would prefer to leave it this way if you don't mind"), but NASA stands firm. Hour 163: Lovell is in, and Borman is out. Eventually, Berry prevails, and both suits come off. Otherwise, Berry recalls in his oral history, "I don't think we would have completed fourteen days in that spacecraft. . . . You've got two guys in spacesuits and they're sitting like this, your leg over in the other guy's lap. It's a really difficult situation."

It could be worse. Try living in bed for three months.

II

THE HORIZONTAL STUFF

What If You Never Got Out of Bed?

Leon M. doesn't appear to have the "right stuff." He has a messy past and lingering debts. His most recent job was as a security guard. These days, Leon spends entire weeks in bed, watching movies and playing video games. Beneath the sweatpants and tattoos, however, there's an astronaut of sorts. Leon's skeleton has been diminishing at about the same rate as an astronaut's in space.

Leon is part of a NASA-funded bed-rest study at the Flight Analogs Research Unit (FARU) at the University of Texas Medical Branch in Galveston. For decades, space agencies around the world have been paying people rather handsomely to lounge around all day and night in their PJs. That's how it was presented to Leon, who heard about the gig on one of Howard Stern's oddball-headline roundups: NASA WILL PAY YOU TO LIE IN BED.

For three months, twenty-four hours a day, Leon does not get up—or even sit up—for anything: not to shower, not to eat, not to use the toilet. Bed rest is an analog, or mimic, of spaceflight in that staying off one's feet causes the same sorts of bodily degradations that weightlessness causes. Most direly, the bones thin and the muscles atrophy. Space agencies study bed-resters to try to understand these changes and figure out how best to counteract them.

Bed-rest studies often assess the helpful (or not) effects of drugs or exercise devices—countermeasures, as they say in aerospace medicine lingo—but the one for which Leon has volunteered is simpler. The researcher is comparing certain changes in men versus women. Leon pauses an episode of *Magnum, P.I.* on the smartphone that he bought on the Internet with his first check. "So basically, yeah, I'm just deteriorating. And they just want to watch it." He reports this as cheerfully as someone else might report a promotion or a good night at the blackjack table. Leon has high cheekbones, longish, springy black hair, and an appealing smile.

The human body is a frugal contractor. It keeps the muscles and skeleton as strong as they need to be, no more and no less. "Use it or lose it" is a basic mantra of the human body. If you take up jogging or gain thirty pounds, your body will strengthen your bones and muscles as needed. Quit jogging or lose the thirty pounds, and your frame will be appropriately downsized. Muscle is regained in a matter of weeks once astronauts return to earth (and bed-resters get out of bed), but bone takes three to six months to recover. Some studies suggest that the skeletons of astronauts on long-duration missions never quite recover, and for this reason it's bone that gets the most study at places like FARU.

The body's foreman on call is a cell called the osteocyte, embedded all through the matrix of the bone. Every time you go for a run or lift a heavy box, you cause minute amounts of damage to your bone. The osteocytes sense this and send in a repair team: osteoclasts to remove the damaged cells, and osteoblasts to patch the holes with fresh ones. The repaving strengthens the bone. This is why bone-jarring exercise like jogging is recommended to beef up the balsa-wood bones of thin, small-boned women of northern European ancestry, whose genetics, postmenopause, will land them on the short list for hip replacement.

Likewise, if you stop jarring and stressing your bones—by going into space, or into a wheelchair or a bed-rest study—this cues the strain-sensing osteoclasts to have bone taken away. The human organism seems to have a penchant for streamlining. Whether it's muscle or bone, the body tries not to spend its resources on functions that aren't serving any purpose.

Tom Lang, a bone expert at the University of California, San Francisco, who has studied astronauts, explained all this to me. He told me that a German doctor named Wolff figured it out in the 1800s by studying X-rays of infants' hips as they transitioned from crawling to walking. "A whole new evolution of bone structure takes place to support the mechanical loads associated with walking," said Lang. "Wolff had the great insight that form follows function." Alas, Wolff did not have the great insight that cancer follows gratuitous X-raying with primitive nineteenth-century X-ray machines.

How bad can it get? If you stay off your feet indefinitely, will your body completely dismantle your skeleton? Can humans become jellyfish by never getting up? They cannot. Paraplegics eventually lose from one-third to one-half of their bone mass in the lower body. Computer modeling done by Dennis Carter and his students at Stanford University suggests that a two-year mission to Mars would have about the same effect on one's skeleton. Would an astronaut returning from Mars run the risk of stepping out of the capsule into Earth gravity and snapping a bone? Carter thinks so. It makes sense, given that extremely osteoporotic women have been known to break a hip (actually, the top of the thighbone where it enters the pelvis) by doing nothing but shifting their weight while standing. They don't fall and break a bone; they break a bone and fall. And these women have typically lost a good deal less than 50 percent of their bone mass.

NASA funded the work that led to Carter's computer models. "But it seems like no one there read our report," he says. "They have this idea that they can send astronauts up and the bone loss will level off in a few months, but the evidence that has come back doesn't support that view. If you look at a two-year mission to Mars, it's kind of a scary prospect."

SOME BED-REST FACILITIES call their volunteers "terranauts." At first I assumed this was done to confer a sense of importance to the pursuit, like calling a janitor a sanitation engineer. But the day-to-day existence of the three-month terranaut bears similarities to that of an astronaut orbiting Earth. Each day begins with wake-up music on the speaker system. (It was Metallica* on the space station this morning; "some Beethoven thing" at FARU.) You spend your time confined to a small room, or cluster of rooms, and if you try to go outside, you are in trouble. Privacy is hard to come by. At FARU, closed-circuit cameras are aimed at the beds, so staff can be sure everyone is staying flat. (Subjects are allowed to pull the curtain that surrounds their bed only when they use the bedpan.) Whiners are not a good fit. Leon says he

---

* Astronauts' families take turns picking it. In the Gemini era, Mission Control would pipe in music, not always in a pleasing manner, as this Gemini VII exchange suggests:

CAP COM [capsule communicator]: . . . How do you like the music?

COMMAND PILOT FRANK BORMAN: We turned it off. We got a little busy there and we turned if off for a while.

CAP COM: Okay. They've got some good Hawaiian stuff coming up to you.

went through an irritable patch at the halfway point of his stay, but that he is "so chipper they didn't notice." In the half hour I've spent with Leon, I have heard only one complaint. It involved the chicken. "It's little squares. I want chicken with a bone and skin on it! Don't give me those cubes."

Leon excuses himself, because the masseuse is coming. Unlike astronauts, bed-resters get a massage every other day to help with the lower back pain that is a common side effect of taking a load off. Obliviously, doctors used to prescribe bed rest for patients with lower back pain. According to a 2003 article in *Joint Bone Spine*,* regardless of what ails you, it is almost always a good idea to get out of bed as soon as you can.

Without the weight of a body compressing it, the spine's curvature lessens and the discs between the vertebrae expand and absorb more water. Astronauts are as much as 2.5 inches taller after about a week in space. (The typical gain is 3 percent of one's height.) Like children, they will "outgrow" their suits if a "growth" spurt has not been factored in.

AARON F. HAS BEEN "head-down" for eight weeks. (The term refers to the 6-degree tilt of the beds. Since weightlessness causes body fluids to shift to the upper part of the body, so must bed rest.) A large fan by his bed is running at top speed, not to cool him but to mask the noises out in the hall. He's been feeling trapped, unable to get away from it. Not helping matters: His roommate Tim is still in his "ambulatory period." He goes head-down in a

---

* An unusual display of syllabic restraint in journal naming. Only *Gut* earns my higher praise. Take note, *American Journal of Orthodontics and Dentofacial Orthopedics, Official Publication of the American Association of Orthodontists, Its Constituent Societies, and the American Board of Orthodontics.*

couple days, but for now he's allowed to pad around the unit in his slippers and sit cross-legged on his bed, which he is doing now.

A kitchen worker pushes a serving cart into the room.

"The high point of my day!" says Tim. He looks genuinely excited over the prospect of hospital food. Aaron accepts his tray without comment. He props himself on one elbow. It is odd to see people reclining for a meal. It's a drab, antiseptic take on a scene from the Arabian Nights, men lounging on pillows, eating with one hand.

Tim takes me on a guided tour of supper, pointing with his fork. "We've got chicken . . ."

I think of Leon. "Diced?"

"Diced, yes. You could almost roll them! And over here are carrot coins, . . ." There is a rapt quality to his speech, as though we were gazing at gold doubloons. ". . . apple slices, milk, two rolls, Jell-O. I really love the food here."

Aaron searches for something positive to say. "It's a good variety." But is struggling. "Then again, it's the same variety. We get a lot of fish—"

"Oh my god." Tim again. "The fish is *amazing*!"

Tim reenlisted after his first stint here, a few years ago. A sign on his wall says WELCOME BACK, 9290 in glitter paint borrowed from the pediatric oncology unit next door.

Before I can stop him, Tim slides off his bed to go ask the kitchen staff whether there is an extra dinner for me.

Aaron is antsy and squirmy, alternately bringing his legs up to form an A-frame under the sheets and then stretching them flat again. Like Leon and others I spoke with, he is here because he's trying to pay off some credit card bills. Bed-rest studies are a modern-day debtor's prison. It's not just the amount of money—$17,000 for three months of service—but the limited opportunities to spend it. For three months, there is no rent to pay, no groceries or gas to buy, no bar tabs, no air fares. A bed-rest stint is a way to force oneself out of bad habits. (Though not entirely effective: Internet shopping has made FARU one of the busiest stops on the local UPS route.)

Tim graduated with a business degree and no money with which to start a company. He moved into a Vipassana ashram because he felt a need to ponder his future and because "they feed you, and it's free!" After much thought and rice, he decided to become an actor. He spent the next four years as "a starving artist, literally," and then he heard about a study here at FARU. When he finished, he returned to acting, joining a New Hampshire theater troupe doing "children's Macbeth," the very thought of which alarms me. When the chance came to re-up at FARU, he took it. These days he's weighing wildly diverging career options: joining the Houston police force, opening a coin-op laundry, enrolling in Navy officer candidate school, starting a landscaping business, and becoming a motivational speaker. He is having, as he puts it, "a quarter-life crisis."

According to FARU manager Joe Neigut, 30 percent of the people who sign on to bed-rest studies say they are doing so not only for money, but to be a part of the space effort. As Leon says, "It's as close as I'm going to get to being an astronaut." At the very least, the association with spaceflight puts a luster on the undertaking. Knowing this, the staff entreat astronauts to write thank-you messages on 8-by-10 glossies. Every now and then, an astronaut drops by to deliver them in person. Aaron got an in-person visit but does not recall the man's name. Tim received an autographed photo of Peggy Whitson. ("A total BAMF* astronaut," he called her.)

---

* I had to look up BAMF on Google. It stands for Bad Ass Motherfucker, but don't tell that to the Berkeley Avenue Mennonite Fellowship or the Builders' Association of Metropolitan Flint.

Tim is back from the kitchen. There is no extra food for me, and that's okay. "Did I miss anything?"

"Yuh," Aaron says. "I moved to the left a bit."

THE BIGGEST SKELETON at Johnson Space Center belongs to John Charles. Charles is 6 feet 7. When he was ten, he knew he wanted to be an astronaut. His skeleton, as though aware of its fate in space, sabotaged Charles's dreams by growing past the astronaut height cutoff. Charles got his Ph.D. in physiology and went to work for NASA. It is his job to do what he can to protect the bodies and bones of astronauts.

Charles and I spoke one recent afternoon in the Lyndon B. Johnson Meeting Room in the public affairs building at Johnson's namesake, the Johnson Space Center. A chaperone from the Public Affairs Office sat quietly in the corner, as though Charles and I might otherwise leap into each other's arms there amid the plaques and signed proclamations of the Johnson era. Charles must put the public affairs people on edge. He is known for speaking his mind freely and sits high enough on the ladder of command not to worry too much about the consequences.

As on Earth, weight-bearing exercise is the best way to hang on to your bone. In zero gravity, of course, you have to create your weight. The problematic and expensive way to do this is to outfit the space station with a rotating room, a huge, inhabitable centrifuge that spins astronauts outward toward the walls, creating artificial gravity. (Keir Dullea can been seen jogging on one in *2001: A Space Odyssey.*) The funky and affordable alternative is to mimic weight by pulling astronauts' bodies down into a treadmill belt as they jog. Typically this involves a harness and bungee cords and much cursing and chafing. It is not tremendously effective. Bone loss researcher Tom Lang says this kind of device pulls exercisers against the belt with about 70 percent of their bodyweight, a scenario that still translates to "massive bone loss."

It's unclear how much exercise helps. "Exercise is probably better than not exercising in space," says Charles, "but we don't know how much better, because we've never done the experiment." No one wants to expose a control group to the sort of bone loss that could result from doing no exercise at all. "If you have hundreds of astronauts who've done different levels, you can pool them into groups and see that this group did slightly less and it had this effect, and this group used a treadmill, not a bike, and it had that effect. But we don't have those large numbers. We have one person that used a bicycle and not a treadmill, one person that used a bike and then changed to a treadmill, and the first is a female in her forties and the second is a male in his sixties. All we can do is a sort of grouped average. The grouped average says that we have countermeasures that still are not protecting astronauts as much as we would like them to be protected." According to Lang, astronauts are coming home from six-month space station stints with 15 to 20 percent less bone than they had when they left.

FARU has lately run a study on vibration as a means of preventing bone loss. Subjects exercised while pulled by elastic cords into a vibrating plate installed at the foot of their bed. It's the same kind of vibrating plate you see advertised on the Internet with promises to build bone and muscle, trim fat, flatten bellies. I was surprised to find them here. So was John Charles. When I asked him about vibration as a bone-loss countermeasure, he said, "It's over with. It's not working." The FARU consent form notes that the investigator has a "relationship" with the vibration machine. He helped invent it.

Carter, too, was surprised to hear about the vibration study. He says the only promising data came from an animal study in which vibration appeared to speed fracture healing. "But in animals that just had low bone mass, it hardly changed the bone mass at all."

Vibration has had an enduring quack appeal. Medical journals from 1905 to 1915 are rife with articles on "vibratory massage" and the many things it cures. Weakened hearts and floating kidneys. Hysterical cramp of the esophagus and catarrh of the inner ear. Deafness, cancer, bad eyesight. And lots and lots of prostate problems. A Dr. Courtney W. Shropshire, writing in 1912, was impressed to note that by means of "a special prostatic applicator, well lubricated, attached to the vibrator, introduced to the rectum" he was "able to empty the seminal vesicles of their secretions." Indeedy. Shropshire's patients returned every other day for treatment, no doubt also developing a relationship with the vibration machine.

Neither Tim nor Aaron is involved in an exercise study. "Me allowing myself to atrophy is going to be the hardest thing I've ever done in my life," says Tim. Before he began the study, Tim was running three to five miles three times a week. He has a countermeasure plan of his own devising. "I heard a story of a POW in Vietnam." He pauses for some Jell-O. The spoon clicks against the glass bowl. "He was locked in a cage." Click-click-click. "Every day he played golf mentally. He improved his golf score by six strokes!" He leans back against his pillow. "So, mentally, I can go on a jog."

Aaron has been pinching off pieces of dinner roll and listening without comment. He turns to face us. "I've been mentally doing squats." He says he has considered suggesting to NASA that they enlist yoga masters or Buddhist monks to teach astronauts how to train their minds to fight the effects of zero gravity. I'm mentally enjoying the image.

The dinner cart returns, and the trays are taken away. The attendant places Tim's glass on his table. "You didn't finish your milk," she says. Food intake is documented as part of the studies.

Students hired to monitor the bed-resters make sure they don't stuff food under their mattresses or behind the ceiling tiles. (Both have happened.)

"You have to eat everything," says Aaron. "They will bring back your little tub of maple syrup and make you drink what's left in it."

PEGGY WHITSON HAS lived through the scenario that worries Dennis Carter and John Charles. In this scenario, astronauts who have been weightless in space for months or years, bone and muscle compromised, find themselves in an emergency situation: enduring the G forces of a crash landing, jumping out of capsule hatches, pulling colleagues to safety. For Whitson, as we learned earlier, it came to pass in 2008. She and two crewmates returning from the International Space Station endured a ballistic reentry and a 10 G landing. Sparks from the landing set the grass afire, and crewmate So-yeon Yi injured her back.

I talked to Whitson* about the incident. The day the interview was scheduled, there were technical problems with the phone system. By the time Whitson's voice came on the line, six of my allotted fifteen minutes were up. I lurched from niceties straight into

* Like all astronaut activities, interviews are exactingly planned and timed. They are like tiny space missions. Whitson's and mine was aborted and rescheduled twice. When the moment finally arrived, my call was relayed via an operator to a booth where Whitson would be sitting. Time passed. "I'm not getting an answer," the operator said. "What time are you scheduled for?" I told her 12:30. "Okay, you're calling early," she said. "I've got 12:28 P.M." You'll hear the NASA TV commentator say things like "The sleep shift is scheduled to start at 1:59 A.M. Central Time. Crew due to awaken at 9:58 A.M. Central Time." Sleeping pills? You betcha.

fire and snapping bones. "Commander, I am a huge admirer. Were you worried that your legs would break when you had to run away from the Soyuz capsule?"

"Nah," said Whitson. She had more pressing concerns. Breathing during the 8 G's of reentry, for instance, and not throwing up in front of Kazakh farmers in the field where they'd landed.

On her first ISS mission, Whitson said, she exercised so much that some of her bones were denser* than they were before she left. Her overall loss was less than 1 percent. "I did so many squats that I actually increased some in my hips." Tom Lang, who has studied the skeletons of ISS astronauts, is not overly reassured by things like this. The returning astronaut's total bone mass can be very similar to what it was before the mission, but that mass is distributed differently. Most of the regrowth takes place in the parts of the bone needed to support walking. But the parts of the hip that would break in a fall were nowhere near where they had been, leaving women like Whitson vulnerable to fractures in their retirement years.

When you fall, the top of your hip—or more specifically, the femoral neck and greater trochanter at the top of your thighbone—takes the brunt of the force in a side-smack manner. That's not the same architecture that gets strengthened when you jog or

---

* You often read that astronauts' skulls get thicker in zero G. I assumed that this was because the extra fluid in the top half of the body plumps the brain, and that the body responds to the increased pressure by thickening the cranium—just as it responds to increased blood pressure by thickening the arteries. "Interesting hypothesis," said NASA physiologist John Charles. Then he told me it's not true that living in space makes astronauts' skulls thicker. Or not literally anyway. Charles says they do routinely develop the "space stupids"—cognitive impairment brought on by "sleep deprivation, over-scheduled timelines, and all the other indignities we heap onto astronauts."

do squats. The parts of the bone that are stressed by walking and everyday activity hold up surprisingly well with age. The body tends to redistribute bone to those areas—at the expense of other structures, including the ones you fall onto. For this reason, some osteoporosis experts feel that fall prevention is a better way to avoid broken hips than is load-bearing exercise.

I asked Tom Lang whether anyone had looked into the possibility of preventing hip fractures by simply thwacking the aged on the sides of their hips a few times a day. Not hard enough to break anything, obviously, but vigorously enough that the impact would stimulate the osteocytes to strengthen the structure. I didn't expect him to say yes. He told me to contact Dennis Carter at Stanford University.

"It was just a concept," said Carter when I called. "We never built it." It didn't thwack, it squeezed. "You'd sit in a lounge chair and have things at your sides squeezing your hips, right at the greater trochanter, where people fall and hit their hips." It seems like a smart idea, but the companies Carter approached wouldn't touch it. Because they thought the hips might break and the ladies would sue? "That, yes. And I think it was just too weird for them."

Is it possible to bolster one's hip bones by doing some type of controlled fall? Here too, I did not expect a yes. Carter told me that a graduate student at the Oregon State University Bone Research Laboratory had looked into this. As part of her thesis, Jane LaRiviere had subjects lie on one side, raise themselves up 4 inches, and then drop onto a wood floor. They did this thirty times in a row, three times a week. At the end of the trial, scans showed a statistically significant, though small, increase in bone density in the femoral neck on that side, as compared with the undropped-upon side. One of LaRiviere's professors, Toby Hayes, felt that if the impacts had been a bit harder and the study lengthier, the results might well have been more impressive.

When you get right down to it, nothing works particularly well. Calcium's a bust. To a certain extent, so is exercise. Bisphosphonates have come under scrutiny for giving some patients necrosis of the jawbone. "The state of the art for countermeasures right now," John Charles allowed, "is the same as it was forty years ago."

The astronauts don't care. "They want to go to Mars," says Charles. "That's what they joined the program for."

WHITSON IS CONFIDENT that someone will come up with a good, safe drug solution by the time a manned Mars mission becomes a reality. A more likely scenario is that genetic testing will by then play a part in astronaut selection. (There's a large hereditary component to bone loss.) Charles envisions NASA recruiting Mars astronauts who are "almost bulletproof—people who never had a kidney stone in their lives, that come with high bone density, good cholesterol numbers, high radiation insensitivity . . ."

The bones of black women are 7 to 24 percent denser, on average, than those of white and Asian women. (I don't have statistics for black men, but presumably they have sturdier bones as well.) I asked Charles whether NASA ought to consider an all-black crew for Mars. "Why not?" he said. "For decades, we had an all-blond, blue-eyed program."

An all–black bear crew would be another way around the bone-loss conundrum. Black bears emerge from their dens after four to seven months in bed with bones as strong as when they turned in. There are researchers who believe that hibernating bears may hold the key to treating and preventing bone loss. I talked to one of them, Seth Donahue, an associate professor of biomedical engineering at Michigan Technological University. Donahue said that hibernating bears' bones do break down, just like bed-resters' and astronauts' bones. What's different is that their bodies take the calcium and other breakdown minerals out of the blood and reapply it to their bones. Otherwise the calcium level in their blood would build to a lethal concentration. Because during those four to seven months, the bears don't get up to go to the bathroom. All the bone minerals that get dumped in the bloodstream as the bones dismantle themselves would stay there, accumulating. "So they've evolved a method to recycle that calcium." And therefore not die. The bone protection is "a lucky consequence."

Donahue and others have been studying the hormones that control bear metabolism to see whether they can identify some component that will help postmenopausal women (and astronauts) grow new bone. They've nominated bear parathyroid hormone. Donahue has a company that makes a synthesized version, injections of which are being tested in rats and eventually, if all goes well, will be tested in postmenopausal women. Even human parathyroid hormone makes women grow bone. It's one of the most effective ways to increase postmenopausal bone density. Unfortunately, high doses make rats grow bone cancers, and thus the Food and Drug Administration limits prescriptions to one year and for women who've already had fractures. Donahue said bear parathyroid hormone doesn't appear to have any adverse side effects, so keep your claws crossed that it pans out.

There's another reason hibernating bears are interesting to NASA. If humans could be made to hibernate, to breathe one-fourth as much oxygen and eat and drink nothing for six months of a two-to-three-year Mars mission, imagine how much less food and oxygen and water one would need to launch. (The less baggage on board a spacecraft, the cheaper it is to launch. Once it reaches the speed needed to escape the pull of Earth's gravity and leaves behind the air drag of Earth's atmosphere, a spacecraft basically coasts to Mars.) Each extra pound of weight launched adds thousands of dollars to the project budget. Science-fiction writers glommed onto the idea decades ago, outfitting fictional spacecraft with high-tech, climate-controlled hibernaculums.

Do space agencies ever discuss human hibernation? They have, and they do. "It never dies," says John Charles. "It just hibernates." Charles puts little stock in the possibility. "Even if it did work, would we really short-supply a crewed vehicle on a three-year mission to Mars? What if the hibernaculum malfunctioned, and everyone woke up? How much food and oxygen do you carry, just in case? And when is that amount sufficiently large that the savings due to hibernation are lost?"

Here's another reason it won't work. Hibernating bears derive all their water and energy from reserves of fat that they build up by bingeing before they den. According to the Bear Center at Washington State University, a small (astronaut-sized) bear gorging on apples and berries consumes up to 40 percent of its body weight each day during this period. That's about 65 pounds of food a day.

Six months of living on nothing but fat—even your own—probably isn't healthy unless your body has somehow adapted to it. Little known fact: Hibernating bears have high "bad" cholesterol levels. (They also have very high "good" cholesterol—which probably explains why heart disease is unknown in bears.)

BED-RESTERS ARE not bears. They have to eat and drink and excrete, and that last one was Tim's undoing. At FARU, B's are to be M'd in bed, and no place else. Using a bedpan while lying flat on one's back is an awkward and unnatural way to "make," as my mother-in-law Jeanne likes to say. Tim sat up, and was caught on film by the camera aimed at his roommate Aaron's bed. (He hadn't drawn the curtain around that side of his bed because Aaron was out of the room.) "I didn't think it would have that much of an

impact," he told me. "But it really threw off the scientific data."*
Tim was asked to leave.

Leon had no trouble with this particular aspect of bed rest. "After the first couple times, it's second nature. And I go . . . *a lot.* I go at least four or five times more than any subject here. By the end of three months, I'll be at around 260. . . ." This is one way bed-resters are different from astronauts. With bed-resters, there are no taboo interview topics.

Including sex. Earlier, Joe Neigut was showing me the shower area, a tiled room the size of a horse stall, outfitted with a waterproof gurney. "So the shower," I said, "is their only . . . *private* time, do you know what I mean?"

"Yes . . ." Joe replied. Then he began talking about the new shower head, which had replaced an industrial sprayer of the type used by restaurant dishwashers. I wasn't sure he did know what I meant, so I asked Leon. Leon confirmed that the shower was "where most of them do it." As with astronauts in orbit, masturbation is not formally addressed in the FARU rules or orientation. Leon, being Leon, asked the unit psychologist. "I mean, if it's something that would throw off the test or something, I wouldn't do it." The psychologist blushed and then gave Leon the go-ahead, leaving the logistics up to him.

In a memoir, astronaut Michael Collins relates a story of a physician back in the Apollo era who recommended regular masturbation on long missions, lest astronauts develop prostate infections. The flight surgeon for Collins's moon mission "decided to ignore that advice," and ignoring seems to have been the basic

---

* How often do research subjects cheat? From skimming the posts on Guinea Pig Zero, I'd say pretty often. "Everyone cracks open their pills to see if they're cornstarch," says one drug study subject on the topic of supposedly blind control groups.

approach to the human sex drive ever since. It's the same way at the Russian space agency. Cosmonaut Alexandr Laveikin told me he too had heard that lengthy abstinence could cause prostate infections, but that the space agency pretends the issue doesn't exist. "It's up to yourself how you will deal with it. But everybody is doing it, everybody understands. It's nothing. My friends ask me, 'How are you making sex in space?' I say, 'By hand!'" As for the logistics: "There are possibilities. And sometimes it happens automatically while you sleep. It's natural." John Charles told me he'd heard about the link between prostate health and "self-stim"—at NASA, there's an abbreviation for everything—but never heard any formal discussion, pro or con, of orbital masturbation.

Or two-party sex, for that matter. Here at FARU, that is covered in the rules, though indirectly. Visitors can't sit or lie down on the beds. "My wife didn't mind," jokes Leon. "That was a plus of me leaving!" I had stopped in to his room again to say good-bye. He's been showing me family photos on his computer.

"I should probably get going. I know you've got . . ."

Leon grins. "Nothing to do?"

12

THE THREE-DOLPHIN CLUB

Mating Without Gravity

Sean Hayes was taking off his wet suit when I called. Hayes is a marine biologist who wrote his dissertation on harbor seal mating strategies. Since floating in water is a useful approximation of floating in zero gravity—useful enough that astronauts rehearse spacewalk duties in a giant pool—and since it is easier to get a seal expert (hell, a *seal*) to expound on weightless sex than it is to get NASA going on the topic, I turned to the marine biologists.

"They're very discreet,"* said Hayes, of earless seals in general (as opposed to the shore-mating, circus-ball-balancing eared variety). Hayes built special equipment to spy on wild harbor seals and still never caught a glimpse of floating pinniped bliss. In its natural habitat, the spotted seal, much like the spaceman, has never been caught in the act. If you want to see how it's done,

---

* You would be too if your foreplay included "creaky door vocalizations" and coming to the surface to "maintain eye contact as they breathe heavily into each other's faces."

you need to put a couple of them in a swimming pool. Hayes sent me a paper written by two Johns Hopkins researchers who did just that.

What the biologists observed confirmed what I had suspected: that when it comes to sexual intercourse, gravity is your friend. "The male spent most of his time grasping the female tightly, attempting to hold on and remain in the coital position," the researchers wrote. He used his teeth as a third hand, biting onto the female's back to help keep the two of them from floating apart.* A photograph shows the blubbery couple on the bottom of the pool, attempting to counteract Newton's Third Law: To every action there is an equal and opposite reaction. Take away or greatly reduce the force of gravity, and thrusting just pushes the object of one's affections away.†

Unlike the spotted seal, astronauts have not been put in a swimming pool for the purposes of figuring out how it's done. Regardless of what the late G. Harry Stine says in his book *Living in Space*:

---

* Further evidence of the difficulties of reduced-gravity sex comes from the sea otter. To help hold the female in place, the male will typically pull the female's head back and grab onto her nose with his teeth. "Our vets have had to do rhinoplasties on some of the females," says Michelle Staedler, sea otter research coordinator at the Monterey Bay Aquarium. (Sex can also be traumatic for the male otter, who endures aerial pecking attacks by seagulls mistaking his erect penis for a novel ocean delicacy.)

† This is no doubt the reason that even Steven "the Hunter" Hunt, the man whose pictures and video feed comprise underwatersex.net, chose to opt out of neutral buoyancy and "drop down about 30 feet to a sand bar" for his "Nude Scuba" encounter with an unnamed "bored, lonely housewife." Says Steve: "Can you imagine all the positions you can do while weightless?" You'll have to, because Steve runs through the same old positions you'd see back in the dive shack, only with unattractive, face-distorting scuba gear.

> Back in the 1980s, some clandestine experiments were conducted very late at night in the neutral buoyancy weightless simulation tank at NASA's George C. Marshall Space Flight Center in Huntsville, Alabama. The experimental results showed that yes, it is indeed possible for humans to copulate in weightlessness. However, they have trouble staying together. The covert researchers discovered that it helped to have a third person to push at the right time in the right place. The anonymous researchers . . . discovered that this is the way dolphins do it. A third dolphin is always present during the mating process. This led to the creation of the space-going equivalent of aviation's Mile High Club known as the Three Dolphin Club.

Stine is best known for writing science fiction, and seemed unable to shake the habit while writing nonfiction. Or did someone at Marshall perhaps start the rumor? I wrote to a public affairs officer there to see if anyone could shed light on the story's origins. Squirreliness ensued: "Hi, Mary. I'm including our historian, Mike Wright, on this email as he can probably fill you in on some historical information about the Neutral Buoyancy Lab. The short answer is, yes, we used to have a Neutral Buoyancy Lab at the Marshall Center, but it was closed (Mike can provide dates) and the work was subsequently done at the Johnson Space Center in Houston." It was as though my email had made no mention of sex or G. Harry Stine.

Based on his dolphin accuracy quotient, Stine is not to be trusted. In the words of America's preeminent dolphin expert, Randall Wells, "Only two dolphins are required for mating." Upon further pestering, Wells noted that a second male sometimes helps corral a female but no helpful coital pushing has been observed. One possible reason a third dolphin isn't called for is that the dolphin's penis is prehensile.* Georgetown University dolphin researcher Janet Mann told me it can "hook into the female" and keep her close for the few seconds the male needs to finish his business. However, it was Mann's feeling that the males needed this advantage not so much because it was hard to stay coupled while floating, but because females usually roll over and try to escape. From what I hear about male astronauts, this is not an issue.

As for the research experiment Stine described, it makes little sense. Why would NASA employees risk losing their jobs when the same "experiment" could be carried out in a backyard swimming pool? And why would you even need a formal experiment? As astronaut Roger Crouch said, in an email, a couple that wanted to have sex in space would simply do what couples on Earth do: "just start out and get better by experience."

As for Stine's claim about participants having "trouble staying together," Crouch was dismissive. "Nothing restricts the use of arms and legs to manipulate or cling to each other. Once one of the participants has attached his or her feet or body firmly"—and here he suggested duct tape if all else failed—"the rest would be up

---

* They can literally grasp things—including, on occasion, people who have paid to swim with the dolphins. "There have been cases in captivity when males . . . have gripped the person around the ankle with their penis," said dolphin researcher Janet Mann. Mann said male dolphins have quietly been eliminated from most of the programs for this reason. If the Web site Sex with Dolphins is to be believed, females do it too. "She suddenly decided to grab my foot with her genital slit," writes the author, going on to explain that females not only have muscular vaginal orifices but can use these muscles to "manipulate objects and carry them." What a boon for the limbless! I wanted to ask Mann what objects dolphins have been seen to carry with their genitals, but she had by this point begun dodging my emails.

to the imagination of the participants. The Kama Sutra couldn't start to cover all the possibilities."

I had written to Crouch about a different sex-in-space Internet hoax—NASA Publication 14-307-1792: a fabricated circa-1989 "post-flight summary" of the results of an exploration, supposedly carried out on shuttle mission STS-75, of "approaches to continued marital relations in the zero G orbital environment." It was the first hoax I've ever come across that cited another hoax—Stine's "similar experiments undertaken in a neutral buoyancy tank."

With "a pneumatic sound-deadening barrier" erected between the decks for privacy, an astronaut couple supposedly tried out ten positions, four of them "natural," and six involving mechanical restraints. Position No. 10 was one of two selected as "most satisfactory": "Each partner gripping the other's head between their thighs." The report concluded with a recommendation to screen future astronaut couples based on "their ability to accept or adapt to the solutions used in runs 3 and 10" and a reference to a forthcoming astronaut sex training video. Incredibly, two authors of space books, over the years, swallowed the bait and presented Document 14-307-1792 as fact in their books. A quick visit to the NASA Web site would have revealed that shuttle mission STS-75 flew in 1996, seven years *after* the "document" appeared, and, P.S., had an all-male crew.

DOZENS OF ASTRONAUTS have flown on coed crews. One shuttle crew included a couple who'd fallen in love during training and tied the knot without telling NASA, just before their flight. It's hard to imagine that all these men and women, without exception, have resisted temptation. Privacy may have been scarce on the Space Shuttle, but not on multimodule space stations like Mir and the International Space Station. Valery Polyakov and the fetching Yelena Kondakova spent five months together on Mir. "We were torturing Valery about whether they had sex," cosmonaut Alexandr Laveikin told me. "He said, 'Don't ask these questions.'" Kondakova is married to cosmonaut Valery Ryumin, which helps explain why Polyakov would have needed to keep his flight suit, or his mouth, zipped. Laveikin shared a Russian saying that seems to have both lost and gained something in translation: "Mystery is the thing where love hides its arrows." Or as space maven James Oberg put it (borrowing an old military aphorism): "Them what says, don't know, and them what knows, don't say."

NASA doesn't specifically address sex in its rules of conduct. Its Astronaut Code of Professional Responsibility includes a vague Boy Scout Oath–style pledge, "We will strive to avoid the appearance of impropriety." To me, that just means, Don't get caught. The ISS Crew Code of Conduct—which is actually part of the U.S. Code of Federal Regulations—is similarly circumspect: "No ISS crewmember shall . . . act in a manner which results in or creates the appearance of: (1) giving undue preferential treatment to any person or entity in the performance of ISS activities . . ." That is one way to look at a sexual dalliance: undue preferential treatment.

In reality, nothing needs to be spelled out or legislated. NASA is funded by taxpayer dollars. Like senators and presidents, astronauts are highly visible public servants. Sexual missteps and other breaches of moral etiquette are not easily forgiven. There would be headlines. Public outrage. Funding cuts. An astronaut knows this. Even if word of a zero-gravity hookup never made it past the ears of NASA, the parties involved would never fly again.

And so, as hard as it is to imagine that no astronaut has had sex in space, it is equally hard to imagine that they have. I tried to explain this to my agent Jay: The years of education and training. The anxiety of not knowing whether there will be another flight.

The extraordinary commitment and devotion to career. There's so much at stake, so much to lose. Jay listened to me, and then he said, "Might be worth it, no?"*

AN ENTIRE FLEDGLING INDUSTRY has been launched on the imaginations of people like my agent. Space Tourism Society president John Spencer envisions an orbiting "super yacht" featuring "Snuggle Tunnels" and a zero-gravity hot tub. Budget Suites America founder Robert Bigelow, now heading up Bigelow Aerospace in Las Vegas, has begun testing and launching inflatable components for a "commercial space station" to be leased out for research, industrial testing, and space vacations and honeymoons.† Bigelow hopes to be open for business in 2015.

In theory, one shouldn't have to wait for Bigelow's hotel rooms or Spencer's superyacht. What fascinates most people about sex in space is not the altitude of the participants but the fact that they're weightless. That being the case, a parabolic flight might do the trick. Though you'd experience it in twenty-second intervals sandwiched between the medically risky intervals where you both weigh twice your usual weight.

---

* This is the same man who, upon being shown a panoramic photograph of a hauntingly beautiful Martian landscape, remarked, "It looks like the outskirts of Las Vegas." Funny he should say that. As I write this, funding efforts are underway for a $1.6 billion Mars World resort in the desert outside Las Vegas.

† Hopefully not based on his earthbound business model. Here are excerpts of TripAdvisor reviews of the Budget Suites America down the road from Bigelow's firm in Las Vegas. ". . . An awful musty odor. The bed didn't have a frame just some box springs setting on the outdated carpet"; ". . . the pool area smelled of urine . . . the water murky"; ". . . air conditioning don't work . . . tv don't work . . . security acts like Gestapo agents."

Since 1993, the Zero G Corporation has been running commercial parabolic flights on a fleet of Boeing 727s. Have any of the weightless also been pantsless? The man I spoke to, who has since left the company and wishes to remain anonymous, said sex on the plane was most decidedly not an option. Zero G had begun contracting with NASA to take college students and schoolteachers up on reduced-gravity flights to promote the space program among students. If the company started letting people have sex in the plane, NASA would be extravagantly disinclined to renew the contract. Besides, the interested couple would need to charter the entire plane, at a cost of $95,000.

I am not the first to have inquired. Someone from the Mile High Club had contacted Zero G "on many occasions" about renting the plane. This is not so much a formal club with bylaws and dues as a Web site where people who've "joined the club" by having sex on an airplane can go to post their stories. If anyone had had weightless sex on a parabolic flight, you'd think this organization would know about it.

"We are unaware of anyone having attempted this feat," said Phil, the man who answers mail sent to the Mile High Club Web site. "If you find what you are looking for, please let us know so we can post it on the site." Phil attached two photographs of a pair of nameless young parachutists having sex during free fall. Their position was fairly conventional—for sex, if not for skydiving: man sitting, woman astride. The one concession to their unusual aerodynamic circumstances was that the man's arms were flung out behind him, for stability. Diverting, but not a particularly good analog for zero gravity. The force of the wind blast against the man's naked backside would have acted like a surface, creating resistance for the pair to push against. I'm curious as to whether the man ended up with a bout of ram-air flatulence, but not especially curious about the sex.

Only pornographers are suitably motivated to take on the expense of chartering an entire plane for the prospect of weightless sex. Playboy has contacted the Zero G Corporation, as did a producer at Girls Gone Wild. "You wouldn't believe how hard they tried and how much they offered," said my contact, of Girls Gone Wild. The producer and crew ended up chartering a plane in Russia, though no one had sex. It's just more shots of girls displaying their unfettered bosoms, this time additionally unfettered by gravity.

Some months later, leafing through a European magazine called *Colors*, I saw a reference to a 1999 porn film called *The Uranus Experiment*, whose producer had apparently chartered a jet for a parabolic flight. "As the plane dived to earth, there was just enough time to film their copulation scene." The star of the film was a Czech actress named Silvia Saint. Could Ms. Saint be the first human being to have had weightless intercourse?

Though Silvia Saint has a healthy presence on the Internet, her email address proved elusive. An acquaintance who writes a popular online sex column suggested reaching out to a well-connected "adult PR person" she knows named Brian Gross. (Because I am *not* an adult, I took delight not only in the name but in the job description, imagining an alternate category of "child PR person" and wishing that some of them worked at NASA.) A glance at Mr. Gross's client endorsements marked him as a man of great versatility, having represented, at one time or another, both ABC News and Booble: The Adult Search Engine. Mr. Gross provided a lead, which led to another, who said that Saint had left the industry five years ago,*

* At the time she retired, Saint had been in more than two hundred pornographic movies. Though one or two have a hint of class (e.g., the Kubrickian-sounding *Mouth Wide Open*), the bulk of the filmography (e.g., *Hot Bods and Tail Pipe #14*, *The Adventures of Pee Man*) suggests that Silvia Saint, at age thirty-three, had earned a rest.

"moved back to the Czech Republic, and dropped off the face of the earth."

Next stop, Berth Milton, the man whose Barcelona company, the Private Media Group, produced *The Uranus Experiment*. Milton, an affable family man with an unplaceable accent, arranged to have downloads of the *Uranus* films (it's a trilogy!) sent to me and promised to help track down Ms. Saint. The plane upon which the historic act had transpired, he said, was part of a fleet of corporate jets, of which Mr. Milton owned a timeshare.

"You asked a corporate jet pilot to fly parabolas?" I said.

"Exactly."

"Had the pilot ever done this before?"

"No." This was surprising information. But Milton went on about the wear and tear on the jet engines, and how the plane was grounded for two days afterward for inspection and maintenance, and so I chose to believe him.

Milton hadn't been there, so he couldn't remember details from the zero-gravity scenes. This was ten years ago, after all, and Private Media was then releasing ten movies a month. He did recall the cameraman, who was notable among his kind for having been, at one time, a cameraman for Ingmar Bergman.

Milton added that he didn't care for Bergman. "He won a lot of awards, but nobody was looking at his movies. He's just depressing. There's no joy."

I mentioned *Fanny and Alexander*.

"Okay, that's probably the only one that you could watch the whole movie. The rest are terrible."

I have to admit that I felt more joy while watching *The Uranus Experiment 1* than I did watching *The Seventh Seal*. The film opens with a cosmonaut sitting naked on an examining table at the Russian space agency. A white adhesive EKG electrode is stuck to his chest like a nicotine patch. It is an odd touch, given that he's there to deliver a semen sample. In the next room, jowly Russian space agency men discuss a top-secret experiment "to find out how zero G affects the sperm production." Cut to a blonde in a snug white lab coat, a test tube dangling from her manicured fingertips. "Hello," she says. "What a beautiful organ you have there."

I fast-forwarded through this scene and the one at NASA (here pronounced *Nassau*) headquarters, wherein we learn how the agency chooses its female interns. (An aerospace degree appears unnecessary.) I stopped fast-forwarding at the point where the action moves to zero gravity. Two orbiting space shuttles, one Russian and one American, have commenced a belly-to-belly docking maneuver. Even the spacecraft are having sex.

The hatch between the two craft is barely opened and the two crews have their flight suits off. Silvia Saint is holding vertical, bobbing up and down as though taking a dip in a mild chop. Hang on. Hold the phone. Her ponytail is hanging down her back, and other things are hanging down her front. Without gravity, there should be no hangy-downy. This wasn't shot in zero G! The actors' lower legs are hidden behind a console; they're just rising up and down on their toes and waving their arms in the air.

A press release for the trilogy, I note, makes reference to just a single shot "in total weightlessness," and it's in *The Uranus Experiment 3*. I get up off the couch to eject No. 2, but I can't just now. An astronaut orgy, led by a Commander Wilson, has gone live on the giant wall screen at Mission Control. It's being broadcast around the world. Scandal and chaos! NASA is shut down. The American president is on the phone. His suit is too big for him and he's working from a cheap motel room. "This is the work of the KGB! I can smell it."

Commander Wilson and Silvia Saint continue to flaunt the NASA Crew Code of Conduct in installment 3. Perhaps it's my imagination, but Commander Wilson appears better endowed than he did in 1 and 2. Could this be the effects of weightlessness? Without gravity pulling the blood down into the lower half of the body, more of it remains in the upper half. Breasts are larger, and anecdotal information suggests penises enjoy the same plumping effect. "I had an erection so intense it was painful," writes astronaut Mike Mullane in *Riding Rockets*. "I could have drilled through kryptonite."

"I have heard others say exactly the opposite," astronaut Roger Crouch told me, craftily leaving his own drill bit out of it. I called upon NASA physiologist John Charles to referee. Charles said that according to Buzz Aldrin, the Mercury and Gemini astronauts reported a definite lack of activity in that region. "They were going to give an award to the first man who demonstrated a response. Though how to prove it?" Charles mused. He sided with Aldrin and Crouch. And John Charles has medical science on his side. The dividing line between the part of the body that gets more fluid in zero gravity and the part that gets less is right around the diaphragm. It's called the hydrostatic indifference point. "The male jumblies are below that point," says Charles, "and so would seem to be drained, not engorged."

This could have posed a challenge for *The Uranus Experiment*'s male cast. But it didn't, because guess what. Nothing was shot in zero gravity. The cameraman simply filmed the ejaculating commander on his back and then flipped the image upside down so he appears to be floating. I happen to know what a "cum shot in total weightlessness" would look like. I know because I've read the 1972 NASA study "Some Flow Properties of Foods in Null Gravity," and those foods included butterscotch pudding and potato soup. The paper includes the dietician's rendition of the zero-gravity cum shot: a demonstration of how a stream of milk "rapidly forms a perfect sphere." Commander Wilson's butterscotch pudding does not do this.

A fond but accusatory email to Berth Milton earned no reply.

• • •

THOUGH A BIOASTRONAUTICS researcher is unlikely to use a hand job to extract a sperm sample—or to preface it with the line "Hello, what a beautiful organ you have there"—the notion of a space agency studying the effects of weightlessness on sperm is a sound one. If the point of manned space exploration is to prepare us for ever-longer missions off Earth, then space agencies will need to fund research on the effects of zero gravity on human reproduction—not intercourse, but its consequences. One legitimate reason for space agencies to be uncomfortable with astronaut sex is that no one knows what biological perils await an embryo conceived in space. Beyond the protection of Earth's atmosphere, cosmic and solar radiation levels rise significantly. Dividing cells are extremely sensitive to irradiation, thus the risk of mutations and miscarriages rises too.

Radiation is a concern even before cells start dividing. There have been official discussions at NASA about whether female astronauts should consider cryopreserving eggs before long flights. One paper suggested lining male astronauts' flight pants with "organ-shielding . . . for the testes." (John Charles says NASA has not embraced the "extraterrestrial codpiece," or not yet anyway.) Studies of the victims of radioactive fallout from atomic bombs in Japan during World War II suggest that short trips into space shouldn't cause infertility. Astronauts returning from six-month missions don't appear to have had difficulties conceiving back on Earth. But radiation risks are cumulative. The longer you're out there, the greater the dangers. That's why astronauts selected for a two-to-three-year Mars mission would likely be, as John Charles puts it, older folks. "They've already had their kids, and they'll be dead naturally before they really develop a whole lot of cancer."

Is mammalian conception even possible in zero gravity? Not

known. In 1988, bull sperm rode a European Space Agency rocket into orbit to see how weightlessness affected their motility. The sperm moved faster and more easily in zero gravity, which seemed to suggest that weightlessness might enhance fertility. Then along came Joseph Tash and his sea urchin splooge. Tash discovered that one of the enzymes that affects sperm motility—the one that tells them to stop wriggling their tails—was activated unusually slowly. In and of itself, not a big deal. But if weightlessness delayed one enzyme's activation, Tash cautioned, it might delay others—including, say, the enzyme that readies the sperm to deposit their DNA packets. Eggs could be tripped up, as well. British sexologist Roy Levin has speculated that, without gravity, it could be difficult or impossible for the ovum to enter and make its way along the fallopian tube.

Why not send some rats into orbit and see what happens? The Soviet space agency did. In 1979, a group of rats was launched in an unmanned biosatellite. After launch, a compartment separator automatically pulled out, allowing male rats to do the opposite. None of the females came back to earth pregnant, though there were signs that conception had taken place. "What the study suggests is that certain early phases go awry," says April Ronca, an obstetrician/gynecologist who studied mammalian pregnancy and birth in zero gravity at NASA Ames before leaving to take a post at Wake Forest University School of Medicine. "Maybe the placenta can't form. Maybe the uterus can't have proper implantation. Any step along the way could be compromised by zero gravity in ways that we haven't foreseen. We know nothing."

Setting aside the radiation dangers, a zero-gravity pregnancy would seem, simply on an intuitive level, to be less problematic. Given that pregnant women are sometimes confined to bed rest—a popular zero-gravity analog, as we've seen—and that fetuses float in fluid (another zero-gravity analog), weightlessness would not, on the face of it, appear to pose a threat to the developing fetus. Ronca sent pregnant rats into space* for the final two weeks of gestation. Two days after landing, the females gave birth. (NASA stopped short of allowing birth in space, largely because of logistics. Someone would have had to build a birthing support for the females, and a nursing structure to keep the babies from floating away from the teat.) Other than some mild vestibular issues, the babies were essentially normal.

What wasn't normal was the birth itself—even though the rats had come down from space by then. Rats who'd spent two weeks in space had fewer, and weaker, uterine contractions. In Ronca's view, this is a dangerous difference. Contractions play an important role in a newborn's adjustment to life outside the womb. The compressions of vaginal birth cause a huge release of stress hormones in the fetus; these are the same fight-or-flight hormones that fuel feats of extreme strength in adults. "This hormonal surge appears to be very important for getting physiological systems moving. All of a sudden a newborn has to breathe on its own, it has to figure out how to suckle from a nipple. If there aren't enough contractions, the hormone release is smaller and the fetus has a harder time." Studies have shown that infants born via planned C-section, with no contractions—as compared to those delivered vaginally—have a higher risk of respiratory dis-

---

* Ronca and her colleagues designed an investigator flight patch that featured a pregnant space shuttle surrounded by baby space shuttles. (Like the astronauts, the scientists involved in a mission traditionally commemorate their projects with sew-on patches.) NASA nixed the patch, even though it allowed a Homer Simpson "Sperm in Space" patch to fly. (The patch shows Homer's head on a sperm tail. The wife of the sperm investigator has a family connection to *Simpsons* creator Matt Groening.) There may be no sex in space, but there is sexism.

tress and high blood pressure, a harder time expelling lung fluids, and delayed neurodevelopment. In other words, stressing an infant appears to be part of nature's plan. (For this reason, Ronca is also not an advocate of water births.)

It surprised me that in thirty-plus years of orbiting science labs, so little work has been done. Is it institutional conservatism? Male squeamishness over obstetrical issues? Ronca thought it was more a case of priority than prudery. "We don't know much about the effects of weightlessness on any of the body's basic systems—bone, muscle, cardiovascular. We know even less about the brain. Reproduction just has not been high on the list."

And now the funding is gone. NASA's life sciences program has been pretty well gutted. I almost wrote "is dead in the water," then caught myself. The last significant NASA mammalian biology study flew aboard Space Shuttle Columbia in 2003. The rats perished along with the crew. There was nothing anyone could do to save them, though the same cannot necessarily be said for the astronauts.

13

WITHERING HEIGHTS

Bailing Out from Space

The Perris SkyVenture vertical wind tunnel is a hurricane in a can. Air rushes at 100-plus miles per hour through the core of a cylindrical building that resembles an air-traffic control tower. It's probably not the tallest building in Perris—a sprawl of malls and tract homes a couple hours out from Los Angeles—but it feels like it. Up near the top, where the controllers would be sitting, a set of doors open onto the column of wind. Customers lean into the air, open their arms and legs as they fall, and are lifted off their feet. It's the sensation of free fall with no danger or rush: skydiving with its balls removed. If it is your first visit, a staff person helps steady you in case you drift upward and panic and bounce off the walls like an air-popped kernel.

Today is Felix Baumgartner's first visit to SkyVenture, but no one is holding on to him. Baumgartner, a photogenic forty-one-year-old Austrian, is a high-profile skydiver and BASE* jumper. You can go on YouTube and watch Baumgartner jump off the outstretched right arm of the enormous Christ statue in Rio de Janeiro or, more prosaically, the roof of the Warsaw Marriott. For most of his jumps he wears a skydiver's jumpsuit. In the Marriott video, he's dressed in business casual. He's done this to pass through the lobby without arousing suspicion, but the impression it gives, as you watch him walk to the edge of the roof in his tie and dress shirt, is that jumping off buildings is just another day on the job for Felix Baumgartner.

This evening finds Baumgartner dressed like an astronaut. He has traveled to Perris this week as part of the Red Bull Stratos Mission. The mission's aims are twofold. I'm mainly interested in the aeromedical side of it. Baumgartner is testing a modified emergency escape suit made by the David Clark Company, makers of spacesuits since the days of the Mercury space program.† Since

---

* BASE stands for Building Antenna (radio tower) Span (bridge) Earth (cliff)—the four dangerously low things they parachute from. According to a 2007 *Journal of Trauma* study, the death and injury rate for BASE jumping is five to eight times that of skydiving. Though it's lower than you'd think: 9 out of 20,850 jumps off Norway's Kjerag Massif over a ten-year span (run of years) have ended in death.

† NASA turned to David Clark because of the company's experience with rubberized fabric. "A spacesuit is a rubberized anthropomorphic bag," says retired Air Force master parachutist and escape-system tester Dan Fulgham. "Well, we didn't have any experience working with rubber bags. We came across the David Clark Company in Worcester, Massachusetts. They were producing twenty gross per month of bras and girdles for Sears, Roebuck." Fulgham has fond memories of driving up to Worcester for meetings and catching glimpses of fit models wandering around in the back. The contract for the Apollo lunar landing suits went to International Latex, which later became Playtex. This didn't get a lot of airplay at the time.

1986, when Space Shuttle Challenger exploded 72 seconds after launch, astronauts have worn pressure suits not just while spacewalking, but during launch, reentry, and landing—the chanciest parts of a flight. Baumgartner will wear it to keep himself alive during a "space dive" from 23 miles (120,000 feet) up. (It's not technically space—space begins at 62 miles—but it's close; atmospheric pressure at that altitude is less than one one-hundredth of what it is at sea level.) The jump—slated for summer or fall 2010 in an undisclosed locale—will provide escape-system engineers with hard-to-come-by information about the behavior of a falling body in a pressurized suit in extremely thin air and the reactions of that body to transonic and supersonic speeds. Because there's so little air resistance up there, Baumgartner is expected to reach 690 miles per hour, rather than the 120-miles-per-hour terminal velocity of a typical free fall at lower altitude. No one has ever bailed out in a spaceflight emergency, and it isn't clear how best to do it safely in all phases of flight.

Baumgartner says he's proud of the contributions he'll be making to safer space travel, but he's primarily interested in breaking records. The current skydiving altitude record is 102,800 feet. That record was also set by a man testing high-altitude survival gear. In 1960, in a project called Excelsior, Air Force captain Joe Kittinger stepped from an open-top steel gondola carried by a helium balloon and skydived, in a partial pressure suit, 19 miles to the ground. He was testing a multistage parachute system. In his oral history transcript on file at the New Mexico Museum of Space History, Kittinger says he broke the sound barrier while free-falling, but he did not carry the equipment needed to make the record official. Thus Baumgartner will likely also make the record books as the first human to reach supersonic speed without a jet or other conveyance.

The Stratos Mission is funded in large part by Baumgartner's corporate sponsor, Red Bull. Sponsoring extreme athletes is Red Bull's way of telling the world that the brand stands not just for caffeinated pop, but for, as the press releases say, "pushing limits" and "making the impossible happen." Teenage boys with little hope of becoming pro skateboarders or record-breaking BASE jumpers can nonetheless drink the drink and feel the feeling. NASA might do well to adopt the Red Bull approach to branding and astronautics. Suddenly the man in the spacesuit is not an underpaid civil servant; he's the ultimate extreme athlete. Red Bull knows how to make space hip.

Baumgartner looks the part. To quote an industrial cutting materials pamphlet I saw not long ago, he has very good bulk and edge-line toughness. He looks like Mark Wahlberg and sounds like Arnold Schwarzenegger, but he's cooler than either. He's in the wind tunnel now, holding facedown in the classic spread-eagle free-fall position. The spacesuit has been pressurized. I count ten charging red bulls. The logos appear vertically on the suit's arms and legs, making some of the bulls appear to be executing a skydiving move called the sit-fly. Baumgartner reaches around to his front to get a feel for the placement of the ripcord. (He can't see it, because the spacesuit prevents him from bending his neck.) Now he straightens his legs, assessing the suit's flexibility. This adds surface area for the wind to push against, and he shoots up ten feet, and then stops, hovering above a group of onlookers like a Thanksgiving parade balloon.

Not since Joe Kittinger's day have escape suits and emergency parachute systems been tested in high-altitude skydives. (It's too expensive. Baumgartner will ascend in a pressurized capsule suspended below a huge—26 million cubic feet—helium balloon.) They probably should be. With so little air resistance, it's hard to control one's body position. Imagine holding your hand in the wind outside a car window at 60 miles per hour. By angling it slightly to present more or less surface to the wind, you can feel obvious shifts in direction and pressure. If the car were traveling 23 miles in the air, you'd feel none of that. It's harder for skydivers—or astronauts or space tourists ejecting at high altitude—to stop a spin, and a poorly designed suit could make the situation worse. Baumgartner will need to free-fall for about 30 seconds before he gains enough speed to generate the wind force needed to control his position—or to benefit from the emergency stabilization chute he'll carry.

The dangers of spinning were explained to me by retired Air Force colonel and master parachutist Dan Fulgham. Fulgham was Joe Kittinger's backup for the record-setting Project Excelsior jump and a veteran escape-system tester for the U.S. Air Force and NASA. During a test of the X-20 "space plane" ejection system, Fulgham went into a flat spin and experienced centrifugal forces so strong that he could not bend his arms to his chest to pull the ripcord. "It was like I was encased in iron," he told me. His chute opened automatically, but he came close to dying even so. Sensors clocked him spinning at 177 revolutions per minute (rpm). "We ran some monkeys on the centrifuge at Wright-Pat," he said, referring to the Wright-Patterson Aerospace Medical Research Laboratories,, "where the force was outward on the head at about 144 rpm. The brain compressed enough into the top of the skull that it separated from the spinal cord. That should have happened to me." He could also have died from redout, wherein blood is spun into the brain with enough force to rupture vessels. Did you see figure skater Mirai Nagasu with a bloody nose at the end of her 2010 Olympic routine? Same sort of thing. Centrifugal force spun the blood in her head outward like water in a salad spinner.

One thing Baumgartner and the Stratos team want to check today is whether the suit allows him to get into "tracking" posture: angled downward with his arms extended Superman-style in front of him. Tracking position causes the skydiver to move laterally as he falls. This is explained to me by Art Thompson, the technical director of the Red Bull Stratos Mission, who is overseeing tonight's tests. Thompson uses a pair of folded reading glasses to demonstrate. By shifting the center of rotation, the tracking position changes a tight, level turntable spin into a larger, slower three-dimensional spiral. Thompson's glasses track out away from his chest and arc around to the left. If that doesn't work, the forces of the spin will trigger the release of a stabilizing chute called a drogue. The drogue will pull Baumgartner's head upright and keep him from spinning into a redout scenario and, hopefully, save his life. (Unless it deploys prematurely, winds around his neck, and chokes him until he passes out, as Joe Kittinger's did in an Excelsior dress rehearsal jump from 76,400 feet.)

There is no way, down on Earth, to simulate free fall in a near vacuum. The Project Excelsior team used to try by dropping anthropomorphic dummies out of high-altitude balloons. The results were worrisome. On a side note, civilians would sometimes be passing through the drop zone and head over to see what was going on. Because the project was operated in secrecy and the recovery teams behaved in a furtive, scurrying manner—and because the dummies had fused fingers and no ears or noses—rumors began to spread that a UFO carrying aliens had crashed in the scrubland outside Roswell,* and that the military was trying to cover it up.

---

* The dummies were realistic enough to fool a group of officers' wives who had gathered for tea at the home of Air Force General Edwin Rawlings. Without warning, a human form thudded to the ground a few hundred feet from the Rawlings' yard. This was followed by Joe Kittinger driving up in a pickup truck and tossing it in the back and speeding away. The women didn't think it was an alien; they thought it was an airman. Later that day Kittinger received a call informing him that Mrs. Rawlings's guests had complained about the careless nature in which the dead "parachutist" was handled.

On one occasion, the "alien" that people were sure they'd seen was Dan Fulgham. Fulgham and Kittinger crashed one Saturday morning as their balloon came down in a field on the outskirts of Roswell. The 800-pound gondola was freed from the balloon too early and began to tumble, coming to a stop on Fulgham's head. When Fulgham took off his helmet, his head swelled so severely that Kittinger was moved to describe his face as "just a big blob." Fulgham was taken to the hospital at Walker Air Force Base, which was staffed in part by civilians. I asked Fulgham if he recalls people pointing and staring as though they'd seen an alien. "I don't know," he said, "because the only way I could see was to put my fingers up and pry my eyelids open." When Kittinger led Fulgham down the steps of a plane to his waiting wife, the woman asked Kittinger where her husband was. "I replied, 'This is your husband,' and she screamed and began to cry," wrote Kittinger in his witness statement in the Air Force publication *The Roswell Report.* I saw photographs of Fulgham taken after the crash. It was weeks before he looked human again.

Thompson thinks the dummy results were misleading and that high-altitude spinning is unlikely to be a serious problem for Baumgartner. I brought up Fulgham's near-lethal spin and Kittinger's drogue-chute cravat. Thompson pointed out that back then people didn't skydive for sport the way they do now. "They weren't used to the idea of controlling body position in flight. There's been so much advancement." This is evident to anyone who's spent time watching the SkyVenture staff hover and dart like hummingbirds.

But astronauts aren't experienced skydivers like these guys. And while Baumgartner will begin his descent at zero miles per hour, jumping from a balloon that's drifting on air currents, a person ejecting from a spacecraft during reentry would be traveling in the neighborhood of 12,000 miles per hour. It's not a neighborhood you'd want to spend any time in.

• • •

THE RED BULL STRATOS MISSION medical director is well qualified for his post. Jon Clark was a high-altitude parachutist for the U.S. Special Forces. He's been a flight surgeon for NASA Space Shuttle crews, and he was involved in the Columbia investigation. (Space Shuttle Columbia disintegrated during reentry in February 2003; a piece of foam insulation had broken off the external tank and knocked a hole in the left wing during launch, damaging the thermal protection that the craft needed to reenter the atmosphere safely.) Clark's team examined the remains of the crew to determine at what point in the disaster's unfolding they had perished and how, and whether anything might have been done to save them.

Clark isn't here in Perris today. I met him more than a year ago, up on Devon Island, where I'd gone for the lunar expedition simulations at the HMP Research Station. I heard him before I saw him. His tent was pitched next to mine, and each evening around eleven, I'd hear the pained exhalations of a middle-aged human trying to get comfortable on hard-frozen ground. The night I finally met Clark, he showed me a PowerPoint presentation about the technologies that air forces and space agencies and, lately, private companies have come up with to keep fliers and astronauts alive when things go wrong. It also covered the things that happen when those technologies fail—as Clark put it, "all the things that can kill ya."

We sat at his desk in the medical tent. No one else was around. A wind turbine outside made a haunted droning sound. At one point, without comment, Clark handed me an STS-107 mission patch, like the one the Columbia astronauts had worn on their suits. I thanked him and set it down on the desk. It seemed like a good time to ask about his work on the Columbia investigation.

I knew from reading the Columbia Crew Survival Investigation Report that the astronauts had not had their visors down when the crew compartment lost pressure. I wondered whether they might have survived had their suits been pressurized and had they been equipped with self-deploying parachutes. The closest thing to a precedent was the crash of Air Force test pilot Bill Weaver. On January 25, 1966, Weaver survived when his SR-71 Blackbird broke up around him while traveling Mach 3.2—more than three times the speed of sound. His pressure suit—and the fact that he was flying at 78,000 feet, where the air is about 3 percent as dense as the air at sea level—protected him from the friction heating and windblast that would, at lower altitudes, handily kill a person moving that fast. Columbia was traveling at Mach 17, but given the negligible density of the atmosphere at 40 miles up, the windblast was about the equivalent of a 400-miles-per-hour blast at sea level. (More on windblast shortly.) It presented what Art Thompson describes as a manageable risk. "It's survivable," said Clark.

But the Columbia astronauts faced crueler threats than windblast and thermal burns. "We had some very unusual injury patterns that were not explainable by anything that we are accustomed to," Clark said. By "we," he meant flight surgeons: people accustomed to brains spun off their stems and limbs snapped by windblast.

"We know how people break apart," Clark continued. "They break on joint lines." Like chicken. Like anyone with bones. "But this wasn't like that. It was like they were severed, but it wasn't from some structure cutting them up." He spoke in a flat, quiet manner that reminded me of Agent Mulder from *The X-Files*. "And it couldn't have been a blast injury, because you have to have an atmosphere to propagate a blast."

ular servings of lard flakes and pregelatinized waxy maize starch? How long could a human being survive on the kinds of foods being dreamed up by military test kitchens? More direly, how long would he *want* to? What does this sort of food do to morale?

Throughout the 1960s, NASA paid lots of people lots and lots of money to answer these questions. Space food R&D contracts were handed out to the Aerospace Medical Research Laboratories (AMRL) at Wright-Patterson Air Force Base and, later, the School of Aerospace Medicine (SAM) at Brooks Air Force Base. The U.S. Army Natick Laboratories drafted the manufacturing requirements, commercial vendors did the cooking, and AMRL and SAM inflicted them on Earth-bound test subjects. Both these bases constructed elaborate space cabin simulators where teams of volunteers were confined for mock spaceflights, some for as long as seventy-two days. Food was often tested at the same time as spacesuits, hygiene regimens, and different cabin atmospheres—including, delightfully, 70 percent helium.

Three times a day, experimental meals would be left by dieticians inside a pretend airlock. Over the years, recruits survived on all manner of processed and regimented aerospace foods: cubes, rods, slurries, bars, powders, and "rehydratables." Dieticians weighed, measured, and analyzed what went in, and they did the same with what came back out. "Stool samples were . . . homogenized, freeze-dried, and analyzed in duplicate," wrote First Lieutenant Keith Smith in a nutritional evaluation of an aerospace diet that included beef stew and chocolate pudding. You had to hope Lieutenant Smith kept his containers straight.

A photograph from this era depicts a pair of men in impossibly cramped conditions, wearing hospital scrubs and belts with some variety of vital-signs monitor. One young man sits hunched on the lower tier of a bunk bed so narrow and thin as to resemble a double-decker ironing board. He holds what appears to be a petit-

I was looking at the Columbia patch. The seven crew members' last names were stitched around the perimeter: MCCOOL RAMON ANDERSON HUSBAND BROWN CLARK CHAWLA. *Clark*. Something clicked in my head. When I had first arrived on Devon Island, I'd heard that the spouse of one of the Columbia astronauts would be here. Laurel Clark was Jon Clark's wife, I now realized. I didn't know whether to say something, or what that something would or should be. The moment passed, and Clark kept talking.

The atmosphere at 40 miles up is too thin for blast waves, but not for shock waves. The investigation team concluded, mostly through a process of elimination, that that's what killed the Columbia astronauts. Clark explained that in breakups at speeds greater than Mach 5—five times the speed of sound, or about 3,400 miles per hour—an obscure shock-wave phenomenon called shock-shock interaction comes into play. When a reentering spacecraft breaks apart, hundreds of pieces—none with the carefully planned aerodynamics of the intact craft—are flying at hypersonic speeds, creating a chaotic web of shock waves. Clark likened them to the bow waves behind a water-skier's boat. At the nodes of these shock waves—the places where they intersect—the forces add together with savage, otherworldly intensity.

"It basically fragmented them," Clark said. "But not everyone. It was very location-specific. We had things that were recovered completely intact." He said one of the searchers who combed the Columbia's 400-mile debris path in Texas found a tonometer, a device that measures intraocular pressure. "It worked."

The wind outside the medical tent had picked up. The turbine made a tortured sound. It was a strange evening. We sat side by side, staring at the slides on Clark's laptop, him narrating and me listening. Occasionally I'd interrupt with a question, but not the ones on my mind. I wanted to ask him how he had coped with learning the details of his wife's death. I wondered why he had chosen to join the investigation. It seemed insensitive to ask. I imagine he got involved for the same reason he's involved in the Red Bull Stratos Mission. He wants to learn everything he can about the things that happen to human bodies when the vehicle in which they are traveling breaks apart at high altitudes and crazy speeds. He wants to apply what he learns to design technologies that can be put in place to protect those bodies, to keep astronauts and space tourists alive, to keep families intact.

It is an extremely complicated challenge. Any spacecraft escape system works for a limited range of altitude and speed. Ejection seats, for instance, will work for the first eight to ten seconds of launch, before Q force—as the interplay of air density and speed-generated wind force is known—builds to a lethal level. An ejection system needs to quickly blast the astronauts far enough away from the craft to keep them from smashing into its appendages or getting caught in the fireball of a catastrophic explosion. The most recent Space Shuttle escape system employed a long pole that crew members would hook onto to slide out away from the craft and clear its wing. Retired aerospace engineer and space historian Terry Sunday points out that this would only work well if the shuttle were flying in stable, straight-and-level flight. "And in that case," says Sunday, "why would you want to leave it?"

To survive the extreme speed and heat of reentry is yet more problematic. The Russian space agency has tested prototypes of an inflatable crew escape pod called a ballute (an amalgam of *balloon* and *parachute*). Heat shielding on the broad forward face of the pod protects the terrified occupant, and the large surface area creates the drag needed to slow the pod to a speed where a multistage parachute system could, if all goes well, lower it safely to Earth. It has never flown all the way from space to the ground. Alternatively, a parachute system could lower an entire capsule or crew cabin to the ground. (Current plans call for NASA's new Orion capsule to be used initially as an ISS escape pod.) The chute would be heavy and costly to launch—and in the case of the Space Shuttle, the process of separating the crew compartment from the rest of the craft presented serious technical challenges. Also, the parachute would need its own heat shielding to keep it from melting during reentry, and this would make deployment trickier.

What about airplane passengers? Is there a way to bail out safely from a jet that's about to crash? Why, other than the weight and expense, don't airlines outfit every seat with a portable oxygen supply and a seat-back parachute? Many reasons. Time for a short primer on windblast and hypoxia.

AT THE HALFWAY POINT of the Beaufort Wind Force Scale, air is traveling 25 to 31 miles per hour. "Umbrella use becomes difficult," states the Beaufort, a tad overdramatically. The scale tops out at 73 to 190 miles per hour—hurricane-force wind. That is all the blow nature can muster. Where the Beaufort leaves off is where windblast studies begin. Windblast isn't weather. The air isn't rushing into you; you are rushing into it—having bailed out or ejected from an imperiled craft.

At the speed of a typical private plane—135 to 180 miles per hour—the effects of windblast are mainly cosmetic. The cheeks are pressed flat against the skull, bestowing a taut, over-face-lifted appearance. I know this both from hideous photographs of me in the SkyVenture wind tunnel and from a 1949 *Aviation Medicine* paper on the effects of high-velocity windblast. In the latter, a man identified as J.L., handsome at 0 miles per hour, appears in a 275-miles-per-hour windblast with his lips blown agape, gums in full view like an agitated, braying camel.

At 350 miles per hour, the cartilage of the nose deforms and the skin of the face starts to flutter. "The waves begin at the corners of the mouth . . . and progress across the face at the rate of about 300 per second to the ear, where they break, causing the ear to wave." Umbrella use is out of the question. At faster speeds this Q force causes deformations that can, as the *Aviation Medicine* paper gingerly phrases it, "exceed the strength of tissue."

Cruising speed for a transcontinental passenger jet is between 500 and 600 miles per hour. Do not bail out. "Fatality," to quote Dan Fulgham, "is pretty much indicated." A windblast of 250 miles per hour will blow an oxygen mask off your face. At 400 miles per hour, windblast will remove a helmet—as it did to Bill Weaver's SR-71 copilot. His visor was blown open and acted like a sail, snapping his head back against the neck ring of his suit and breaking his neck. At 500 miles per hour, "ram air" blasts down your windpipe with enough force to rupture various elements of your pulmonary system. An unnamed test pilot mentioned in a paper by John Paul Stapp ejected at more than 600 miles per hour. The windblast pried open his epiglottis and inflated his stomach like a pool toy. (This worked to his advantage, as he had ejected over water. "The estimated three liters of air in the stomach substituted as flotation gear, which he was in no condition to inflate," wrote Stapp.)

At supersonic speeds, your body would be coping with the kind of Q force that used to regularly shake experimental jets to pieces. Dan Fulgham has heard about pilots who ejected at 600-plus miles per hour. "Ejection seats back then had metal wings on each side of the head to keep it from flopping around," he told me. "When they did autopsies they found the brains had just been emulsified because of the tremendous vibration of the head between those steel plates." Whenever they can, fighter pilots stay with a crippled jet until they can slow it down, reducing the Q load and raising their odds of survival. Red Bull has cause to be nervous about Baumgartner. He could be vibrated to death inside his suit as he approaches or surpasses the speed of sound.

The immediate and dire consequence of plunging into thin air is lack of oxygen. At 35,000 feet, a human being has 30 to 60 seconds of "useful consciousness." You'd definitely want to be first in line at the emergency exit. I can tell you what it's like to wander out to the edges of useful consciousness. As a prerequisite for the weightless flight I undertook in chapter 5, the engineering students and I took a NASA aerospace physiology seminar that included a hypoxia (not enough oxygen) demonstration inside the Johnson Space Center altitude chamber. By pumping air out of a sealed chamber, technicians can simulate the atmosphere at any altitude, all the way to near-total vacuum—a big box of outer space. Space agency personnel use these chambers to test spacesuits and other equipment that will be exposed to the vacuum of space.

After about a minute with our oxygen masks off at 25,000 feet—where one has two to five minutes of useful consciousness—we were asked to complete a list of mental tasks. One question read, "Subtract 20 from the year you were born." I felt fine, but I remember puzzling over it, feeling utterly stumped, and moving on. One of the last questions was: "What does NASA stand for?" I obviously know this, but my answer reads, "N."

More than useful consciousness you would need luck, given that 400 other panicked passengers are bailing out with you, creating a significant danger of tangled parachute lines and canopies. But it would be possible to survive, provided you stay with the plane until it slows to a more survivable speed. You might experience pain, but nothing major. At higher altitudes, as air pressure drops, air inside the body's own chambers tries to unbutton its shorts and expand. A pocket of gas inside an unfilled tooth cavity may press painfully on nerves. Same sort of thing happens to air in the sinus cavities—particularly if they're congested. Even gas dissolved in the cerebrospinal fluid inside the brain's ventricles tries to expand. If I'd had a hole in my skull, my fellow students in the altitude chamber could have watched my brain bulge out of it.* The gas expansion you are most likely to notice is in your digestive tract. At 25,000 feet, air in the stomach, for example, expands threefold. "Go ahead and fire it off," our instructor told us, as if eleven male college students needed an invitation.

BAUMGARTNER IS TAKING a break. He's slumped in a chair with his helmet in his lap, sipping water. (Perris SkyVenture doesn't stock Red Bull.) Art Thompson, the project technical director, is in a good mood. The suit is working well, and Baumgartner feels comfortable in it. (As comfortable as anyone ever feels in a spacesuit. As spacesuit historian Harold McMann put it, "It's not a nice place to be. It's not even a nice place to visit.")

As you read this, there's a good chance Felix Baumgartner will have completed his history-making jump. As I write this, I don't how it turns out. I am cautiously optimistic. Skydiving from extremely high altitude is risky, but probably not as risky as Baumgartner's more typical occupation—jumping from extremely *low* altitude. If something starts to go wrong during a space dive, you have five minutes to figure out how to remedy it. On a BASE jump, you don't have five *seconds*. BASE jumpers don't carry reserve chutes, as there's no time to deploy them. "That's

---

* A proven fact. In 1941, scientists at the Mayo Foundation's Laboratory for Research in Aviation Medicine convinced a woman with a postsurgical hole in her cranium to sit inside their altitude chamber while they took her up to 28,000 feet. The patient (and never was the term *patient* more apt) was positioned in front of a centimeter scale while the researchers, like golf caddies, planted a small triangular flag in the hole to mark the spot. At 28,000 feet, the little flag on her brain had risen a full centimeter.

why they don't tend to have a long . . ." Thompson searches for the right word.

"Life span?" I offer.

"Career."

Thompson says he isn't worried. "Eventually, most BASE jumpers get complacent, but Felix is really anal about what he does. That's what keeps him alive."

Brave and anal: the ideal space explorer. Though you don't find "anal" on any of those lists of recommended astronaut attributes. NASA doesn't really use words like *anal*. Unless they have to.

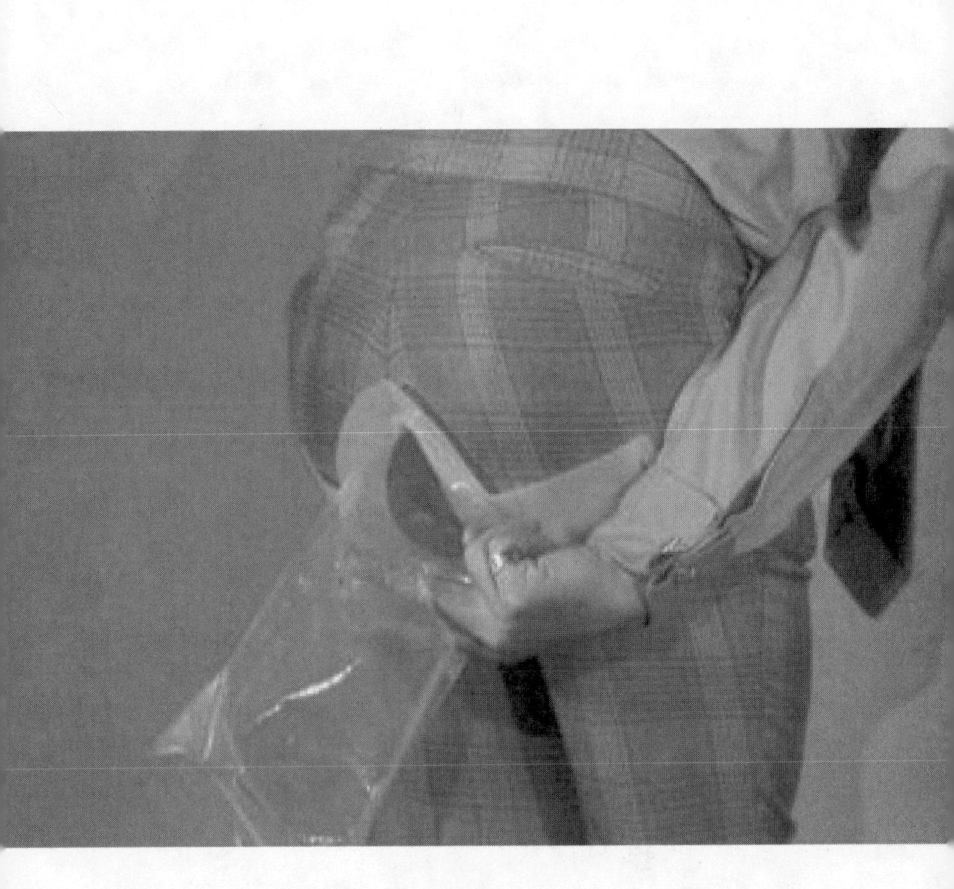

14

SEPARATION ANXIETY

The Continuing Saga of
Zero-Gravity Elimination

It is probably not the first time that a bunch of guys got together and installed a closed-circuit video camera in a toilet bowl at a government agency. It is surely the first time it has happened with the blessings and financial backing of the agency. And that the monitor has been mounted right there in the bathroom, angled for the viewing ease of the person on the toilet.

On the wall to the sitter's left is a small plastic sign:

Positional Trainer
Sit Down on Trainer Seat and Spread Buttocks

The Johnson Space Center "potty cam," as it is more casually known, is an astronaut training aid. It provides a vivid, arresting perspective on something you've had intimate contact with all your life but never really seen. Perhaps not unlike viewing one's home planet from space for the first time. Positioning is critical because the opening to a Space Shuttle toilet is 4 inches across, as opposed to the 18-inch maw we are accustomed to on Earth. Jim Broyan, a waste-water engineer who designs toilets and other amenities for NASA astronauts, is showing me around. Broyan has a reedy build and an angular face. He peers at his visitor over the top of a pair of wire-frame glasses. He possesses a stealthy deadpan wit and is, I imagine, tremendous fun to work with.

"The camera enables you to see if your butt, your . . ." Broyan pauses in search of a better word: not more polite, just more precise. ". . . *anus* lines up with the center." Without gravity, you can't reliably gauge your position by feel. You are not really sitting on the seat. You are hovering in close proximity. The tendency, says Broyan, is to touch down too far back. Then your angle of approach is off, and you sully the back of the transport tube and plug some of the air holes that encircle the rim. Bad, bad move. Space toilets operate like shop vacs; "contributions," to use Broyan's word, are guided along, or "entrained," by flowing air rather than by water and gravity, two things in short-to-nonexistent supply in an orbiting spacecraft. Plugged air holes can disable the toilet. Additionally, if you gum up the holes, it is then your responsibility to clean them out—a task Broyan understates as "arduous."

The room with the potty cam is a working bathroom, complete with sink and paper towel dispenser, but it functions primarily as a classroom. Every astronaut must be potty-trained by Scott Weinstein, who has just joined us. Weinstein is also in charge of galley training—how to eat in space. His is a one-of-a-kind teaching position: taking the most skilled, credentialed, highest-achieving individuals in the world and putting them back in nursery school. Everything these men and women learned as toddlers—how to cross a room, how to use a spoon, how to sit on a toilet—must be relearned for space.

Scott is a big guy, 6 feet 5 and not without some cushioning. He has young kids, and it is easy to picture him with them—on his lap, on his back, climbing him like a play structure. Though he has a background in waste management, he spent seven years elsewhere in NASA, plotting rocket trajectories. Eventually Weinstein realized he wanted to work with people. I imagine he's very good at what he does. His genial, matter-of-fact nature makes it easy to sit down with him and have a talk about things one doesn't routinely talk about.

That is more important than you think. Zero-gravity excretion is not entirely a joking matter. The simple act of urination can, without gravity, become a medical emergency requiring catheterization and embarrassing radio consults with flight surgeons. "The urge to go is different in space," says Weinstein. There is no early warning system as there is on Earth. Gravity causes liquid waste to accumulate on the floor of the bladder. As the bladder fills, stretch receptors are stimulated, alerting the bladder's owner to the growing volume and delivering an incrementally more insistent signal to go. In zero gravity, the urine doesn't collect at the bottom of the bladder. Surface tension causes it to adhere to the walls all around the organ. Only when the bladder is almost completely full do the sides begin to stretch and trigger the urge. And by then the bladder may be so full that it's pressing the urethra shut. Weinstein counsels astronauts to schedule regular toilet visits even if they don't feel the urge. "And it's the same with BMs," he adds. "You don't get that same sensation."

Broyan and Weinstein have offered to let me try the Positional Trainer. Weinstein reaches over to the wall and flips a switch that illuminates the inside of the bowl. Because once you sit down, you are blocking the light from the ceiling fixtures. "So," says Weinstein. "You're going to try to align yourself, flip on the light, see how you did."

I ask him whether the astronauts are observing *while* they go, or before they start.

Broyan appears quietly stricken. "You can't defecate on that toilet." He glances at Weinstein, the briefest of glances yet unmistakable in its message: *Oh my god oh my god she was gonna crap on the camera.*

I wasn't, honestly.

Weinstein, ever genial: "Well, technically you *can*, but then Crew Systems has to come in and clean it up."

"It's not a working toilet, Mary," says Broyan, just to be sure I'm clear.

It has happened just once, a hit-and-run. "It was before my time," says Weinstein. "If I'd been here, I'd have been pulling security tapes." He wishes me good luck. The two of them leave and shut the door.

Imagine stumbling upon an especially rank porno channel, and then realizing it's you on screen. My brain elects to reinterpret the image: *See the funny puppet? Watch his mouth. What's he saying? He's saying, "Ooooo-aaaah-oooooo."*

When Weinstein and Broyan return, Weinstein says he doubts that many of the astronauts use the potty cam. "I get the sense most of them don't want to see themselves." Weinstein provides an alternate positioning tactic, "the two-joint method." The distance between the anus and the front of the seat should equal the distance between the tip of the middle finger and its big knuckle.

Along the same wall as the Positional Trainer is a fully appointed and functioning Space Shuttle commode. It looks less like a toilet than a high-tech, top-loading washing machine. Though the device itself is a high-fidelity version of the one on board the shuttle, the experience is not. There is gravity down here at Johnson Space Center, and that makes all the difference. Gravity facilitates what is known in aerospace waste collection circles as "separation." In weightlessness, fecal matter never becomes heavy enough to break away and drop down and venture forth on its own. The space toilet's air flow is more than an alternate flushing method. It facilitates the Holy Grail of zero-gravity elimination: good separation. Air drag serves to pull the material away from its source.

A separation strategy courtesy of Weinstein: spread the cheeks. That way, there is less contact between the body and the "bolus" (another in the waste engineer's vast arsenal of euphemisms)—and therefore less surface tension to be broken. The newest seat is designed to function as a "cheek spreader" to facilitate a cleaner break.

A more sensible arrangement might be to adopt the posture favored by much of the rest of the world—and by the human excretory system itself. "The squat tends to spread the cheeks," says Don Rethke, a senior engineer at Hamilton Sundstrand, the contractor on many of the NASA waste collection systems over the years. Rethke suggested to NASA that they add a set of foot restraints higher up, to accommodate those who wish to approximate the squatting posture in zero gravity. No go. When it comes to the astronauts' creature comforts, familiarity wins out over practicality. A kitchen table makes little sense without gravity, but all long-duration spacecraft have them. Crews want to sit around the kitchen table at the end of the day to eat and talk and feel normal and forget for a moment that they're hurtling utterly alone through the blackness of a deadly vacuum. In the aftermath of Apollo, where there were fecal bags rather than toilets, bathroom facilities became a charged topic. "When the astronauts came back, they physically and psychologically wanted a sit-down commode," says Rethke.

Understandable. The fecal bag is a clear plastic sack, similar to a vomit bag in its size, holding capacity, and ability to inspire dread and revulsion.* A molded adhesive ring at the top of the bag was designed for the average curvature of an astronaut's cheeks. It rarely fit. The adhesive pulled hairs. Worse, without gravity or air flow or anything else to foster separation, the astronaut was obliged to employ his finger. Each bag had a small inset pocket near the top, called a "finger cot."

The fun didn't stop there. Before he could roll up and seal the bag to trap the offending monster, the crew member was further burdened with tearing open a small packet of germicide, squeezing the contents into the bag, and manually kneading the germicide through the feces. Failure to do so would allow fecal bacteria to do their bacterial thing, digesting the waste and expelling the gas that, inside your gut, would become your own gas. Since a sealed plastic fecal bag cannot fart, it could, without the germicide, eventually burst. "The test of a good friend was to hand the bag to your crewmate and have him get that germicide completely mushed in with the fecal material," Gemini and Apollo astronaut Jim Lovell told me. "I'd go, 'Here, Frank, I'm busy.'"

Given the complexity of the chore, "escapees," as free-floating fecal material is known in astronautical circles, plagued the crews. Below is an excerpt from the Apollo 10 mission transcript, starring Mission Commander Thomas Stafford, Lunar Module Pilot Gene Cernan, and Command Module Pilot John Young, orbiting the moon 200,000-plus miles from the nearest bathroom.

---

* Still, it could have been worse. Also under consideration for the Apollo crews: the "defecation glove." Here the astronaut would reach around and crap in his own palm, then peel back the glove, much as dog owners use a plastic newspaper sleeve to pick up and dispose of dog feces. Then there was the Chinese Finger, a bag that would clamp onto a bolus as you pulled on the end. The name Chinese Finger refers to the cheap party toy of the same name—and possibly to the astronaut's response to the device.

CERNAN: . . . You know once you get out of lunar orbit, you can do a lot of things. You can power down . . . And what's happening is—

STAFFORD: Oh—who did it?

YOUNG: Who did what?

CERNAN: What?

STAFFORD: Who did it? [laughter]

CERNAN: Where did that come from?

STAFFORD: Give me a napkin quick. There's a turd floating through the air.

YOUNG: I didn't do it. It ain't one of mine.

CERNAN: I don't think it's one of mine.

STAFFORD: Mine was a little more sticky than that. Throw that away.

YOUNG: God almighty.

[And again eight minutes later, while discussing the timing of a waste-water dump.]

YOUNG: Did they say we could do it anytime?

CERNAN: They said on 135. They told us that— Here's another goddam turd. What's the matter with you guys? Here, give me a—

YOUNG/STAFFORD: [laughter] . . .

STAFFORD: It was just floating around?

CERNAN: Yes.

STAFFORD: [laughter] Mine was stickier than that.

YOUNG: Mine was too. It hit that bag—

CERNAN: [laughter] I don't know whose that is. I can neither claim it nor disclaim it. [laughter]

YOUNG: What the hell is going on here?

Broyan showed me a circa-1970 photograph of a NASA employee demonstrating the Apollo fecal bag. The man is dressed in plaid trousers and a mustard-hued shirt with cufflinked sleeves. Like so many photographs from the 1970s, it has surely caused its subject lasting embarrassment. This one more so than most. The man is bending over, his rear protruding toward the camera. A fecal bag adheres to the seat of the trousers. The first two fingers of his right hand are inside the finger cot, poised like open scissors. The last finger is adorned with a wide silver pinkie ring. Though his face is hidden, there is, says Broyan, "speculation" as to his identity. Broyan included the photograph in the history section of the first draft of a recent engineering journal paper he wrote. His superiors asked him to take it out. The feeling was that it was "not the best view of NASA."

Here is Broyan's summary of the astronauts' feedback on the Gemini-Apollo fecal bag system, as presented in that same paper. Clearly not all crew members embraced the scenario with the jollity of Young, Stafford, and Cernan.

> The fecal bag system was marginally functional and was described as very "distasteful" by the crew. The bag was considered difficult to position. Defecation was difficult to perform without the crew soiling themselves, clothing, and the cabin. The bags provided no odor control in the small capsule and the odor was prominent. Due to the difficulty of use, up to 45 minutes per defecation was required by each crew member,* causing fecal odors to be

---

* Because the astronauts' time was rigidly scheduled and because bowel movements generally can't be, crew members were forced into conversations like this one, in the Apollo 15 mission transcript, between Commander Dave Scott and Lunar Module Pilot James Irwin.

> SCOTT: Al, why don't you and I switch off here when . . .
> IRWIN: I'd like to take a crap if I can work it in, Dave.
> SCOTT: Okay.
> IRWIN: Tell me when.

> present for substantial portions of the crew's day. Dislike of the fecal bags was so great that some crew continued to use . . . medication to minimize defecation during the mission.

The Gemini-Apollo urine bags were less odious, but not very much so. Especially when they burst, as Jim Lovell's did during Gemini VII. Lovell, quoted in astronaut Gene Cernan's memoir, described the mission as "like spending two weeks in a latrine." Hamilton Sundstrand suit and toilet engineer Tom Chase neatly summed up the sentiment among engineers and NASA brass at the end of Apollo: "We have to do better."

NASA's first zero-gravity toilet was a hands-on load-and-remove-your-own-bag model designed to facilitate specimen collection* during the medical fact-gathering missions of Skylab. It

---

* Astronaut specimens from the Skylab and Apollo eras are still around, in freezers on the top floor of a windowless high-security building at Houston's Johnson Space Center—the one that houses NASA's collection of (nonbiological) moon rocks. "I am not sure what our inventory of excreta from Apollo is right now," John Charles told me. "Forty years of freezing, with occasional thaws due to power outages during hurricanes, may have reduced them to mere vestiges of their former glory." They were there as of 1996, because planetary geologist Ralph Harvey stumbled onto them when he got lost taking a group of VIPs on a tour. "Back then all the doors opened to the same code," he recalls. "I opened this one door and it was almost like the scene from *Raiders of the Lost Ark*. There were these rows of long, low freezers. They all had a little light on them that's blinking, and a temperature readout, and a piece of tape with the astronaut's name. I'm like, *Shit, they stored the astronauts in here!* and I quickly got the people out. I found out later that was where they stored the astronaut feces and urine." Harvey can't recall the room number. "You have to stumble onto it, that's the only way you can find it. It's like Narnia."

was built into the wall. In the years that followed, to accommodate the psychological and vestibular needs of the crews, NASA engineers and designers began building rooms and labs with a more consistent Earth-gravity-based orientation: tables on "floors" and lighting on "ceilings."

Space Shuttle toilets have always been mounted on the floor, but you would not call them normal. The original shuttle toilet bowl featured a set of 1,200 rpm Waring blender blades positioned a brief 6 inches below the sitter's anatomy. The macerator would pulp the feces and tissue—meaning, if all went well, the paper, not the scrotal, variety—and fling it to the sides of a holding tank. "It was kind of pasted there like papier-mâché," says Rethke. Problems developed when the material in the holding tank was exposed to the cold, dry vacuum of space. (Freeze-drying was a way to sterilize it.) Now it didn't stick together as well. The papier had lost its mâché. When the next astronaut switched on the macerator, tiny bits of fecal wasp nest that lined the walls of the tank would break off and get batted around by the blades, turning to dust that escaped into the cabin of the spacecraft.

Here's how bad it got, as reported in NASA Contractor Report 3943: "Reportedly, astronauts aboard the current STS mission (41-F) have resorted to use of Apollo-style adhesive bags. On previous missions, clouds of fecal dust generated by the zero-gravity toilet have caused some astronauts to stop eating in order that they reduce their needs to use the facility." The same report elsewhere pointed out that fecal dust was not merely disgusting, but could result in "an unhealthy growth of *E. coli* bacteria in the mouth," as used to happen on board submarines plagued by sewage vapor "blowback."

The macerator has long since disappeared, but escapees still occasionally plague the crews. The culprit these days is a phenomenon you will read about in space agency waste collection papers and, one hopes, nowhere else: "fecal popcorning." Broyan gamely elaborates: "Because everything else is frozen, the material that's going in, depending on how hard the stool is, has a tendency to bounce off the walls. You've seen the old air-pop popcorn machines? There's an air flow in there and it's kind of circulating. That material's just floating around in the air stream, and it tends to come back up the tube." Howdy, doody.

Fecal popcorning is the reason Space Shuttle toilets were equipped with rearview mirrors. "We ask them to take a look back there as they shut that slider," Broyan says, "in case there's a piece that's on its way up the tube." Fecal popcorning is the gateway phenomenon to fecal decapitation. You do not want fecal decapitation taking place aboard your ship. If a crew member closes the sliding gate at the top of the toilet transport tube just as a popcorning piece is crowning, the slider gate may decapitate it on its way shut. This is a heinous scenario for two reasons. Any material smeared on the top side of the slider is sharing the cabin along with the crew, and, quoting Broyan, "they're going to smell it." Also, the smearage on the underside will freeze-dry the slider gate shut. Now the toilet's out of order, and everyone has to use the shuttle's contingency fecal waste collection system: the Apollo bag. If you're the boob responsible, you are in for some blowback from your crewmates.

THERE IS NO WAY to anticipate a phenomenon like fecal popcorning. Some things you can't know until you get into orbit. That's why toilets, like everything else that flies in space, get hauled up on a parabolic flight for testing. In this case, the testing poses unique challenges.

Along these lines. Late yesterday afternoon, I got the idea that I wanted to try out the Space Shuttle training toilet. I was already scheduled to meet Broyan and Weinstein and my escort from the public affairs office at noon the next day. *Nine A.M., absolute latest I can do,* said my colon. I called Gayle Frere, my public affairs escort, to try to explain my dilemma and reschedule for first thing in the morning. I caught her at her grandson's graduation, where she had to yell over the noise. I pictured her husband turning away from the festivities to ask what was going on. I imagined Gayle shouting into his ear. *It's that writer. She wants to crap in the shuttle toilet!* I apologized and quickly hung up.

My meandering point being that to schedule an evacuation even within a matter of hours can be awkward. Imagine trying to do so on cue within a twenty-second window of weightlessness. Retired NASA food scientist Charles Bourland was once on board a parabolic flight with a group of engineers testing a zero-gravity-toilet prototype. The toilet had a partial screen set up around it, but Bourland could see the man. "It was number two," he told me. "He was all primed to do his thing but couldn't deliver at the appropriate time. There was a lot of joking and yelled words of encouragement," though not from Bourland, who was fighting motion sickness while testing and sampling seventy-two new Skylab foods, including creamed peas and beef hash, and did not need any additional inducement to throw up.

Some of the testing done in weightlessness has been of a more exploratory nature. "As queer as it sounds, if you want to manage what comes out the back end, you gotta understand what it's doing," said Hamilton Sundstrand engineer Tom Chase, whom I ran into on a simulated moon expedition in the Arctic. Chase was wearing his spacesuit hat that week rather than his toilet hat, but he was game to chat shit. "For instance . . ." Chase began drawing on a pad of Hamilton Sundstrand graph paper balanced on his knee. "Without gravity to pull things straight, they tend to curl as they're coming out."* This was documented by NASA and Hamilton Sundstrand toilet engineers that day in a series of 16-mm films. Thanks to this work, aerospace waste collection systems engineers are not only aware of the curl, they know its range of curvature and most likely direction (backward). They know that the softer ones, up to a point, curl more. Why would they need to know all this? Because the curl can gum up the top of the transfer tube and compromise your air flow.

The films featured both male and female volunteers, the latter consisting of, said Chase, "some gals in the nurses' corps." The footage was classified as limited distribution but, according to Hamilton Sundstrand folklore, regularly traveled beyond its prescribed limits. Pretty much "anybody with a buddy in waste management design" saw them, said one of Chase's colleagues. "They were very, very popular, those films."

Eventually someone who saw the shit also saw the potential for it to hit the fan. "You can imagine the reaction," said Chase—*What if someone does a FOIA on these!* (FOIA stands for Freedom of Information Act, whereby journalists and the public can request copies of unclassified government documents.) The films were destroyed. Chase waxed melancholy about their demise. He is part of the team that had been working on toilets for lunar missions. "It's unfortunate because we were going through this phase here where it would be highly useful to us."

Don Rethke said that the far trickier engineering problems—and thus the bulk of the footage—involved urination. For one thing, liquid tends to adhere to the body in space. "When grav-

---

* Rethke called this the "orange peel effect." The term also refers to a defect in a spray-painted surface, most typically the finish on a car. Either way, the auto body guy owes you an apology.

ity goes away," says Rethke, "surface tension is the next physical force." Even on a human hair, surface tension makes liquids cling. Rethke said that people with longer hair can, in zero gravity, hold two to three liters of water in their hair. NASA needed to know the extent to which pubic hair was compromising female "velocity potential." (Scott Weinstein helpfully describes this as how easy it is to "write your name in the snow.")

Chase began sketching again. "You don't just urinate and get a perfect cylindrical outflow, if you've ever kind of observed what's going on. With gals, there's more in the way of getting a pure stream." I.e., labia and pubic hair. And a weakened stream tends to break apart and form floating blobs. Then Chase told me something quite stunning. He said he'd known women who, while out hiking or backpacking, are "able to take their pants down to their ankles and kind of lean back against a tree and just by moving things around a little bit, getting some room there, be able to fire away and direct it." There was a silence while I contemplated this new and life-changing information. Chase went on. "I'm telling you, women can pee harder than men. But you got to be willing to manipulate the anatomy. There's just some ladies who are more comfortable exploring what is possible than other ladies."

No kind of lady, regardless of comfort level, wants an audience of male toilet engineers and their cronies. Eventually the nurses got wind of what was happening and refused to participate in any more filming. Hamilton Sundstrand was forced to get creative. "One of the guys had a really hairy stomach," said Chase, and here he leaned back in his chair and stuck out his belly. "If he went like this . . ." He placed a palm on either side of his stomach and pushed in toward his belly button, such that it was possible to imagine a vertical fold appearing in the flesh beneath his shirt. ". . . he got about the right look. So in zero G they could spray him with ersatz [urine] solution and film it and they could understand about the droplet formation." Chase released his gut. "That's good thinkin'."

THERE ARE OTHER WAYS to test a zero-gravity toilet. "At NASA Ames Research Center, we have undertaken the task of developing human fecal simulants," writes Kanapathipillai "Wiggy" Wignarajah in a 2006 technical paper. Wignarajah is surely the most sophisticated thinker in this realm, but he is not the first. Others before him—in, for instance, the commercial diaper industry—have employed brownie mix, peanut butter, pumpkin pie filling, and mashed potatoes. Wignarajah pooh-poohs these efforts, as none of these substances comes close to approximating, as he puts it, "how human feces will behave"—i.e., its water-holding properties and its rheology. *Rheology*, in food science, refers to the study of consistency. Consistency is determined by things like viscosity and elasticity. Food technologists have special equipment designed specifically to measure these things, and if they are smart, they will not lend them out to anyone at NASA Ames.

A simulant made from refried beans gets respectable scores from Wignarajah. Though the protein content is too high and thus the water-holding properties are off, the beans are said to look and behave so much like human stool that future visits to the tacqueria have, in my mind anyway, been forevermore altered. The bean-based simulant designers hail from "Umpqua," and by this I assume Wignarajah means Umpqua Community College and not the Umpqua Bank or the Umpqua Indian Tribe.

The NASA Ames simulant blew the Umpqua dump out of the water. The recipe features eight ingredients, including miso, peanut oil, psyllium, cellulose, and "dried coarsely ground vegetable matter." It may not taste as good as the Umpqua simulant but is in every other respect a vastly superior product. The main ingredient is the fecal bacteria *E. coli*, accounting for—as it does in real human feces—30 percent of the weight of the material. I don't know whether the Ames toilet division has colonies of fecal bacteria on site—other than the ones inside the gut of every living employee—or whether they are procured by mail order. Wignarajah did not answer my email.

The one feature lacking in the Ames simulant was fecal odor. To be sure future toilets' odor control measures are living up to expectations, Wignarajah plans to add malodorous compounds to the Ames simulant. Which leads one to wonder, Why bother with a simulant? If they need something that smells like the real thing, why not use the real thing? They do, but only at the very end. "Final testing can be completed with limited experiments on real human feces." So powerful is the taboo against contact with human excrement that NASA researchers have, in days past, run simulations with monkey or dog feces playing the role.

ON THE FRONT of Broyan's polo shirt is a patch from International Space Station Assembly Mission ULF2. The design incorporates various facets of the ISS toilet, arranged inside an oval toilet seat. A slogan reads, Proud to Be of Service.

Broyan has good reason to be proud, as do Weinstein, Chase, Rethke, Wignarajah, and everyone they work with. A successful zero-gravity toilet is a subtle finessing of engineering, materials science, physiology, psychology, and etiquette. As with Wiggy's simulants, if just one element is missing, things don't come out right. And few other technical failures have the power to so reliably and drastically compromise a crew's well-being.

It's possible the elimination issue has had even deeper ramifications. I interviewed a retired Air Force colonel named Dan Fulgham, who had been involved in the selection of the first Mercury astronauts. Colonel Fulgham told me the excretion conundrum was the main reason female pilots weren't considered.* "We knew women were as good as men. We had female pilots all during World War II. They could fly fighters. They could fly bombers." But they couldn't use a condom-ended in-suit urine collection device. "The collection of body waste was a real issue logistically." (The adult diaper was apparently not on anyone's radar screen.)† "We were under the gun to get this thing underway," Fulgham recalled. "So we said, 'Let's limit the amount of concerns we have.'"

If you read *The Mercury 13: The Untold Story of 13 American Women and the Dream of Space Flight*, you'll see that the women pilots had other things working against them. Like Vice President Lyndon Johnson, who, rather than signing a letter to the director of NASA urging him to let female fighter pilots apply to become astronauts, wrote "Let's stop this now!" across the bottom.

As mission lengths grew long enough to require a fecal strategy and crews grew to two-person, the female problem persisted.

---

* It's also the main reason the Russians *did* select females—for animal flights anyway. Training male dogs to urinate into a collection device proved extremely difficult, because cramped conditions in the capsule kept them from assuming their natural posture—lifting a leg.

† According to the Diaper Evolution Time Line on disposablediaper.net, the adult diaper debuted in 1987 (in Japan). Though the general disposable diaper concept dates back to 1942. The inventor was a Swedish company—not, as you sometimes hear, NASA. Skimming the time line, it does sometimes sound as though NASA were involved. There are vacuum-dry diapers, pulpless diapers, diapers with flexible closing systems and "reduced chassis and elastic ears." NASA's adult diapers are COTS—a "commercial off the shelf" product. The current one is a product called Absorbencies. It is hard to imagine a worse name for a diaper, except possibly NASA's previous commercial off-the-shelf adult diaper, Rejoice.

"The issue of privacy had been a big factor in NASA's reluctance to include women as astronauts," writes former NASA psychiatrist Patricia Santy of the Apollo-Gemini era. In *Choosing the Right Stuff,* Santy cites the development of the private space bathroom—"probably more than any other reason"—as the motivating factor behind NASA's decision to allow female astronauts.

Were toilets a *reason* to exclude women, or an excuse? You would think that the passage of federal prohibitions on gender-based hiring discrimination would have been a more powerful impetus than a toilet door. The irony is that female astronauts are the more practical choice for spaceflight. On average, they weigh less, breathe less, and need to drink and eat less than men. Which means less oxygen, water, and food have to be launched.

Rather than keeping launch costs down by flying smaller, more compact humans, NASA chose to fly smaller, more compact pot roast and sandwiches and cake. Rarely has anything so cute been so loathed.

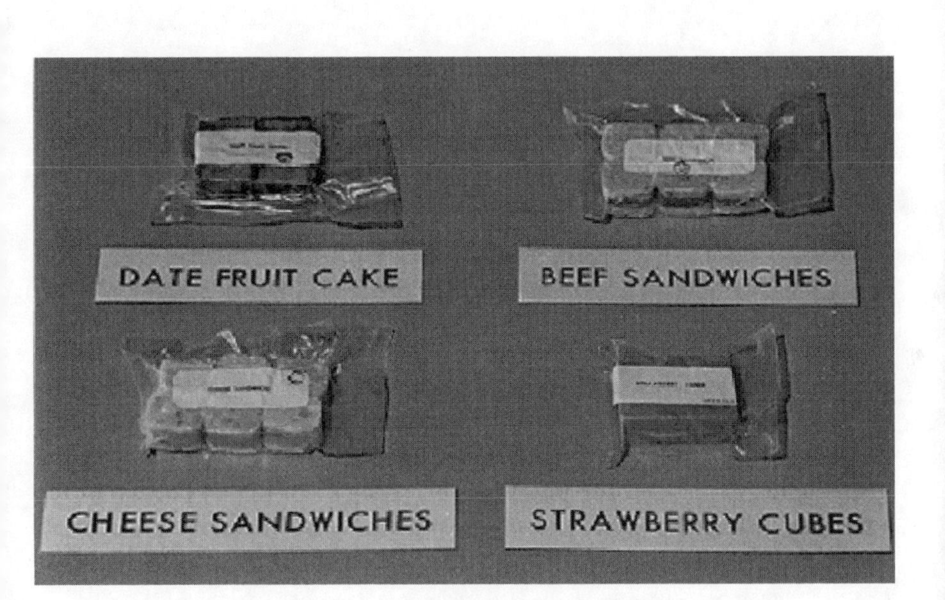

DISCOMFORT FOOD

When Veterinarians Make Dinner, and Other Tales of Woe from Aerospace Test Kitchens

On March 23, 1965, a corned beef sandwich from Wolfie's delicatessen was launched into space. This particular branch of Wolfie's was in Cocoa Beach, Florida, not far from the Kennedy Space Center. Astronaut Wally Schirra ordered it to-go and drove it back to Kennedy, where he convinced astronaut John Young to smuggle it on board the Gemini III capsule and surprise his crewmate Gus Grissom. Two hours into the five-hour-long flight, that is what Young did. The moment did not go entirely as envisioned.

GRISSOM: Where did that come from?

YOUNG: I brought it with me. Let's see how it tastes. Smells, doesn't it?

GRISSOM: Yes, [and] it's breaking up. I'm going to stick it in my pocket.

YOUNG: It was a thought, anyway.

GRISSOM: Yep.

The "corned beef sandwich incident" became ammunition for NASA detractors at congressional budget hearings later that year. In the *Congressional Record* for July 12, 1965, one Senator Morse, pushing for a 50 percent reduction to the proposed $5 billion NASA budget, said Young had "made a mockery" of the entire Gemini science program, with its carefully measured intakes and outputs. Someone else asked NASA administrator James Webb how he could expect to control a multibillion-dollar budget if he could not control two astronauts. Young was given a formal reprimand.

The contraband Wolfie's sandwich violated no less than sixteen of the formal manufacturing requirements for "Beef Sandwiches, Dehydrated (Bite-sized)." The requirements cover six pages and are set forth in the ominous phrasing of biblical commandments. ("There shall be no . . . damp or soggy areas." "The coating shall not chip or flake.") Moreover, the Wolfie's sandwich exhibited Defect #102 ("foreign odor, e.g., rancid") and Defect #153 ("breaks when handled"), among dozens of others but hopefully excluding Defect #151, "visible bone, shell or hard tendonous material."

Food to eat in a space capsule must be the opposite of a Wolfie's deli sandwich. It must be lightweight. Every extra pound that NASA launches into space costs thousands of dollars in fuel needed to lift it into orbit. It must be compact. The Gemini III capsule was no bigger than the interior of a sports car. Because of the strict size and weight limits, space food technologists were preoccupied with "caloric density": packing the most nutrition and energy into the smallest volume of food. (Polar explorers, facing similar constraints and caloric demands but lacking government research budgets, pack sticks of butter.) Even bacon had to be squeezed under a hydraulic press and made more compact (and renamed the Bacon Square).

Compressed food not only took up less stowage—which is how children and aircraft designers say "storage"—space, it was less likely to crumble. To the spacecraft engineer, crumbs were more than a housekeeping issue. A crumb in zero gravity does not drop to the floor where it can be ignored and ground into the flooring until the janitor comes around. It floats. It can drift behind a control panel or into an eye. That's why Grissom stashed the corned beef sandwich when he saw it was falling apart.

Unlike a Wolfie's sandwich, a sandwich cube can be eaten in a single bite. Even a piece of toast will drop no crumbs if you are able to pop the whole thing into your mouth. Which you can do when your toast, as Young and Grissom's did, takes the form of a Toasted Bread Cube. As an extra margin of safety, crumbs were held in check by an edible coating. ("Chill fat-coated toast pieces until they congeal . . . ," goes the recipe.)

The aerospace feeding teams—some Air Force, some Army, some commercial—devoted considerable effort to perfecting the coatings for their food cubes. One technical report outlines a Goldilocksian progression of formulas. Formula 5 was too sticky. Formula 8 cracked in a vacuum. But Formula 11 (melted lard, milk protein, Knox gelatin, cornstarch, sucrose) was thought to be just right. Except by those who had to eat it. "Leaves a bad taste in your mouth and coating on the roof of your mouth," Jim Lovell complained to Mission Control during Gemini VII.

IT IS ONE THING to craft a lacquered sandwich cube that weighs less than 3.1 grams and resists fragmentation "when the sandwich is dropped from 18 inches onto a hard surface." It is another to make this the sort of food a man will happily, healthily eat for weeks at a time. The missions of the Mercury and Gemini programs were, with one or two exceptions, of short duration. You can live on just about anything for a day or a week. But NASA had set its sights on lunar missions up to two weeks long. They needed to know: What happens to the digestive health of a man who consumes reg-

four in his left hand, and a plastic bag containing four more layered cubes in his lap: dinner. A piece of tubing is taped to his nose. His roommate wears black Clark Kent glasses and a communications headset and sits at the kind of console that looked futuristic in 1965 and now looks Star Trek campy. The caption unhelpfully reads: "Space food personnel, 1965 to 1969." Perhaps the writer had tried something more informative—"Testing the effects of miniature sandwiches on heart and breathing rates"—but could find no way to phrase it without compromising Air Force dignity.

Many of the shots are Before photos, luckless smiling airmen posed on the threshold of the SAM test chamber alongside dietitian May O'Hara before they step inside and she closes the hatch. O'Hara looks exactly as you imagine an Air Force dietician to look—neither over- nor underweight, well coiffed and nice-looking, though unlikely to have a profound effect on the heart rate and oxygen uptake of young Air Force recruits. O'Hara was a good Egg Bite. In a military news service article, she voices concern over the acceptability of the various space foods "day after day for 30 days or more."

She seemed to be the lone voice of reason. Though cube foods were getting tepid ratings, their developers pressed on enthusiastically, relentlessly, hydraulically. They could not see that foods that require you to rehydrate them with your own saliva—by holding them "in the mouth for 10 seconds"—might be a spirit dampener on a week-long flight. And they were. On mission after mission, sandwich cubes were, says retired NASA food scientist Charles Bourland, "some of the things that routinely came back." (He means they were still on board after landing, not that they were regurgitated. I think.)

I telephoned O'Hara at her home in Texas, just after lunch on a weekday afternoon. She is in her seventies now. I asked her what she'd eaten. It was a dietician's lunch, and a dietician's answer, laid out like a cafeteria menu: "Grilled beef and cheese sandwich, grapes, and fruit punch." I asked May whether the SAM simulator subjects often quit the studies early or busted out of the airlock to make a midnight run for Whataburger. They did not. "They were all just as cooperative as they could be," said May. For one thing, she explained, they'd just come out of basic training. The prospect of a month with no physical demands more strenuous than chewing had a certain appeal. Plus, in exchange for volunteering, they were given their choice of Air Force assignment, rather than simply being sent someplace.

Over at the AMRL simulator, the volunteers were paid undergrads from nearby Dayton University. Perhaps because they were paid, or because Dayton was a Catholic school, these men too were compliant and generally well behaved. Though missing Communion* occasionally became an issue. One volunteer became so agitated that the scientists broke protocol and summoned a priest, who gave Holy Communion over closed-circuit TV and microphone. Into the pass-through port was placed a small portion of wine and a single Communion wafer, whose palatability probably scored on a par with more typical chamber fare.

---

* Religious observations are even tougher in a real spacecraft. Launch weight limitations forced Buzz Aldrin to pack a "tiny Host" and thimble-sized wine chalice for his DIY Communion on the moon. Zero gravity and a ninety-minute orbital day created so many questions for Muslim astronauts that a "Guideline of Performing Ibadah at the International Space Station" was drafted. Rather than require Muslim astronauts to pray five times during each ninety-minute orbit of Earth, the guidelines allowed them to go by the twenty-four-hour cycle of the launch location. Wipes ("not less than 3 pieces") could be used for preprayer cleansing. And since the orbiting Muslim who began his prayer while facing Mecca was likely, by prayer's end, to be mooning Mecca, provisions were made allowing him to simply face the Earth or "wherever." Lastly, instead of lowering the face to the ground, a trying maneuver in zero gravity, prostrating oneself could be approximated by "bringing down the chin closer to the knee," "using the eye lid as an indicator of the changing of posture" or—in the vein of "wherever"—simply "imagining" the sequence of movements.

One test diet scored even lower than the cubed foods. "It was milk shakes in the morning, lunch, and supper. And the next day, it was milk shakes in the morning, lunch, and supper," says John Brown, the officer who had been in charge of the AMRL space cabin simulator. On a scale of 1 to 9, volunteers who lived on them for thirty days gave the food an average score of 3 (dislike moderately). Brown told me 3 probably meant 1: "The subjects filled out their forms telling you what you wanted to hear." One subject confided to Brown that he and his fellow volunteers had been regularly dumping portions of their formula under the cabin flooring. Despite the diet's unpopularity, the researchers evaluated no less than twenty-four different commercial and experimental liquid diet formulas. I once read an Air Force technical report that lists the desired attributes of edible paper: "Tasteless, flexible, and tenacious." It's how I imagine some of these space food guys.

Meanwhile, over at SAM, Norman Heidelbaugh was testing a liquid diet of his own devising. An Air Force press release called it the "eggnog diet." May O'Hara described it as "sort of a powdered Ensure." "That was really not acceptable," she said with uncharacteristic bite to her words. Heidelbaugh himself seemed to leave a bad taste in people's mouths.

Though it appeared that the science of nutrition was attracting a unique breed of gustatory sadist, other forces were at work here. It was the mid-sixties. Americans were enraptured by convenience and the space-age technologies that bestowed it. Women were going back to work, and they had less time to cook and keep house. A meal in a stick or a pouch was both a novelty and a welcomed time-saver.

That was the mindset that propelled one of the AMRL's least popular liquid diets into a long and lucrative career as Carnation Instant Breakfast. The Space Food Stick also began life as a military washout. What the Air Force called "rod-shaped food for high-altitude feeding" was originally intended as food that could be poked through the port of a pressure suit helmet. "We couldn't get them stiff enough," O'Hara told me. So Pillsbury took back its rods and went commercial with them. Bourland says they occasionally went up with the astronauts simply as an onboard snack—sometimes under the name Nutrient-Defined Food Sticks and other times as Caramel Sticks, fooling no one.

Even the companies who made food sticks and breakfast drinks didn't expect the American family to eat nothing else. I have reason to believe that a cabal of extreme nutritionists was influencing thought at NASA. These were men who referred to coffee as a "two-carbon compound." Who wrote entire textbook chapters on "topping strategies." Here is MIT nutritionist Nevin S. Scrimshaw defending the liquid formula diet at the Conference on Nutrition in Space and Related Waste Problems in 1964: "Persons with other worthwhile and challenging things to fill their time do not necessarily require bits to hold in their mouth and chew or a variety of foods in order to be productive and to have high morale." Scrimshaw boasted of having fed his MIT subjects liquid formula dinners for two months with no complaint. The Gemini astronauts narrowly escaped a fate worse than cubes. "We are hoping, in the Gemini program," said NASA man Edward Michel at that same conference, "to go to some type of formula diet. . . . We will use it during preflight, during the flight, and for a 2-week period post-flight."

Scrimshaw was wrong. People *do* "necessarily require bits to hold in their mouth and chew." Put them on liquid diets and they crave solid food. I spent just one morning on the Mercury-era tube diet, and I did. The astronauts no longer eat tubed food, but military pilots do, when they're in the middle of a mission and can't stop to unwrap a sandwich. Vicki Loveridge, a helpful and congenial food technologist with the Combat Feeding Directorate at U.S. Army Natick, said the formulation and technology have changed little since the Mercury era. Loveridge invited me to Natick. ("Dan Nattress will be making Apple Pie in the tubes on the morning of the 21st.") I couldn't go, but she was kind enough to send me a sampler box. They look like my stepdaughter Lily's tubes of oil paint.

Tube eating is a uniquely disquieting experience. It requires bypassing the two quality control systems available to the human organism: looking and sniffing. Bourland told me the astronauts hated the tubes for precisely this reason: "Because they could not see or smell what they were eating." Also unnerving is the texture, or "mouthfeel," to use a food technology coinage. When a label says Sloppy Joe, you expect some Joe. The Natick version has no discernible ground-beef qualities. It's puréed. All tubed food is, because, as Charles Bourland put it, "the texture is limited to the orifice of the tube." The very first space food was essentially baby food. But even babies get to eat off spoons. Mercury astronauts had to suckle theirs from an aluminum orifice. It wasn't heroic at all. Or, as it turned out, necessary. A spoon and an open container will work fine in zero gravity as long as the food possesses, to quote the adorable May O'Hara, "stick-to-it-ive-ness or whatever." If it's thick and moist enough, surface tension will keep it from drifting off.

The Sloppy Joe tasted like frozen enchilada sauce. The Natick vegetarian entrée—which someone, obviously at a loss, had simply labeled "Vegetarian"—was another vaguely spicy tomatoey purée. Being a Mercury astronaut must have been like being trapped in the sauces aisle of a very small grocery store. But the Natick applesauce—identical in formulation to John Glenn's history-making applesauce tube*—was A-okay.

---

* The first food consumed by a NASA astronaut, but not the first food in space. The Soviets won this space race, too. Glenn's applesauce lost out to Laika's powdered meat and breadcrumb gelatin and the unnamed snack of Yuri Gagarin (in the words of Elena, the Gagarin Museum archivist, "Some say soup, some say purée. For sure there was something in the tube!").

Partly, I imagine, because it's familiar. You expect applesauce to be puréed. One of the problems with the early space foods was their strangeness. When you're hurtling through space in a cold, cramped, sterile can, you want something comforting and familiar. Space cuisine appealed to the American public as a novelty, but astronauts had had enough novelty for several lifetimes.

FROM TIME TO TIME, there was talk among the astronauts that it might be nice to have a drink with dinner. Beer is a no-fly, because without gravity, carbonation bubbles don't rise to the surface. "You just get a foamy froth," says Bourland. He says Coke spent $450,000 developing a zero-gravity dispenser, only to be undone by biology. Since bubbles also don't rise to the top of a stomach, the astronauts had trouble burping. "Often a burp is accompanied by a liquid spray," Bourland adds.

Bourland was in charge of a short-lived effort to serve wine with meals on board Skylab. University of California oenologists steered him toward sherry, because it's heated during production, and thus keeps better. It's the pasteurized orange juice of the wine kingdom. Bottles aren't allowed in space, for safety reasons, so it was decided that the sherry, a Paul Masson cream sherry, would be packaged in plastic pouches inside pudding cans. Further limiting the already limited appeal of cream sherry.

The sherry cans, like any other new technology bound for space, were taken up on a parabolic flight for zero-gravity testing. Though the packaging worked fine, no one on board that day left with much enthusiasm for the product. A heavy sherry smell quickly saturated the cabin, compounding the more standard nauseating attributes of a parabolic flight. "As soon as you opened it," recalls Bourland, "you'd see people grabbing for their barf bags."

Nonetheless, Bourland filled out a government purchase order for several cases of Paul Masson. Just before the sherry went into the packaging, someone mentioned it in an interview and letters from teetotaling taxpayers began arriving at NASA. And so, after having spent God knows how much money on the packaging, requisitioning, and testing of canned cream sherry, NASA scrapped the whole endeavor.

Had it flown, the Skylab sherry would not have been the first alcoholic drink requisitioned by a government as rations for a mission of national service. British Navy rations included rum until 1970. From 1802 to 1832, U.S. military rations included one gill—a little over two shots—of rum, brandy, or whisky with the daily allotment of beef and bread. Every hundred rations, the soldiers were also given soap and a pound and a half of candles. The latter could be used for lighting, barter, or, were you the tidy sort, melted down and used to coat your beef sandwiches.

NUTRITIONISTS WERE NOT entirely to blame for the inhumanity of early space food. Charles Bourland alerted me to something I'd overlooked: the abbreviation "USAF VC" after liquid diet promulgator Norman Heidelbaugh's name. Heidelbaugh was a member of the Air Force Veterinary Corps. So was Robert Flentge, one of the editors of *Manufacturing Requirements of Food for Aerospace Feeding*, a 229-page handbook for preparers of astronaut foods. "A lot of the food science guys were military veterinarians," Bourland told me. Dating back to the Aerobee monkey launches and Colonel Stapp's work with the deceleration sleds, the Air Force has had colonies of test animals and, by necessity, veterinarians (or, for those who felt six syllables weren't enough,"bioastronautics support veterinarians.") According to the 1962 article "The Sky's the Limit for USAF Veterinarians!" their responsibilities came to include "testing and formulating foodstuffs"—first animal and eventually astronautical. Bad news for space crews.

Veterinarians in charge of feeding research animals or livestock were concerned with three things: cost, ease of use, and avoiding health problems. Whether the monkeys or cows liked the food didn't much enter into it. This goes a long way toward explaining butterscotch formula diets and Compressed Corn Flake and Peanut Cream Cubes. It's what happens when veterinarians make dinner. Recalls Bourland, "The vets would say, 'When I feed animals, I just mix up a bag of feed and take it out there and they get everything they need. Why can't we do that with astronauts?' "

Sometimes they did. Witness Norman Heidelbaugh's 1967 technical report, "A Method to Manufacture Pelletized Formula Foods in Small Quantities." Heidelbaugh made Astronaut Chow! The top two ingredients, by weight, were Coffee-mate "coffee whitener" and dextrose/maltose, casting doubt on the vet's claim that the human pellets were "highly palatable." Again, deliciousness was not among this man's overriding concerns. Weight and volume were. By those criteria, Heidelbaugh had a winner: "Caloric density would be sufficient to provide 2600 kcal [2.6 million calories] from approximately 37 cubic inches of food."

Heidelbaugh's space-saving methods sound extreme, but only until you read the solution proposed in 1964 by Samuel Lepkovsky, professor of poultry husbandry at the University of California, Berkeley. "If it were possible to find suitable astronauts who are obese," Lepkovsky begins, seemingly unaware that he is nuts.*

---

* Sorry, I mean innovative. That is the adjective used by the author of Lepkovsky's 1985 UC Berkeley obituary. Here we learn that Lepkovsky coauthored the first atlas of the chicken brain and isolated riboflavin from "several hundred thousand gallons of milk." In what little spare time remained, he enjoyed dancing and amateur stock-market analysis, no doubt reaping great gains in dairy futures.

"An obese person with 20 kilograms of fat . . . carries reserves of 184,000 calories. This would provide over 2900 calories daily for 90 days." In other words: Think of the rocket fuel that could be saved by not launching *any food at all*!

Starving your astronauts for the duration of the mission would resolve another early NASA concern: waste management. Not only was the act of using a fecal bag powerfully objectionable, but the end product stank and took up precious cabin space. "What the astronauts wanted to do is to just be able to take a pill and not eat," says Bourland. "They talked about it all the time." The food scientists tried but failed to make it happen. The astronauts' fall-back solution was to skip meals, a deprivation made bearable by the knowledge of what awaited them inside the meal pouches.

Jim Lovell and Frank Borman would be stuck in the Gemini VII capsule for fourteen days. Fasting was no longer a viable waste management strategy. (Almost though: "Frank went, I think, nine days without having to go to the bathroom," says Lovell in his NASA oral history transcript. At which point Borman announced, "Jim, this is it." And Lovell replied, "Frank, you only have five more days left to go here!") The new imperative at NASA was to develop food that was not only lightweight and compact, but also "low-residue." "On the short missions of Mercury and Gemini," wrote Borman in his memoir, "a bowel movement was rare."

Cue the simulated astronauts again. Technical Report AMRL 66-147, "Effects of Experimental Diets and Simulated Space Conditions on the Nature of Human Waste," details the fourteen very trying days of four men who served as digestive stand-ins for Lovell and Borman in the AMRL simulator. The first diet tested was the infamous all-cube diet: little sandwich cubes, "meat bites," and miniature desserts. It was like dolls were running the kitchen.

The cubes were a digestive fiasco. The coating had been modified, with palm kernel oil used in place of lard. The palm oil waltzed through the gut largely undigested, giving the young airmen steatorrhea, and you and me a new vocabulary word. (Steatorrhea is fatty stool, as opposed to diarrhea, which is watery stool.) The steatorrhea created, to quote the *San Antonio Express*,* "gastrointestinal effects which were incompatible with efficient performance in an orbiting vehicle." The reporter was being coy, but the technical paper spelled it out. Oily stools are foul-smelling and messy. Official descriptor number 3—"mushy but not liquid"—was the one most commonly applied by the subjects (whose day-to-day miseries were amplified by the task of inspecting and scoring their own waste). The report didn't mention anal leakage, but I will. If you have oil in your stools—be it from Olestra or from space food cube coatings—some of it may ooze out. When you have one pair of underpants for a two-week spaceflight, anal leakage is not your pal.

Also tested was one of the liquid diets: forty-two days of milkshakes. The thinking was that a liquid diet would cut down on both the volume of solid waste generated by the men, as well as their "defecation discharge frequency." If you drink it, the thinking probably went, you'll pee it. Not so. Because of all the dissolved fibers in the drinks, "daily mass" (forgive me, Father) sometimes increased significantly, in one case more than doubling.

IRONICALLY, IF YOU wanted to minimize an astronaut's "residue," you could have fed him exactly what he wanted: a steak. Animal

---

* The space simulator diet tests were big news in San Antonio, Brooks Air Force Base's home town. In addition to the *Express* story, the *San Antonio Light* ran a story. The advertisement that ran alongside was for Blue Cross/Blue Shield, then the nation's leading insurer. The tag line, and I will send you a copy if you don't believe me, reads, "Come on, San Antonio! Let's all go No. 1!"

protein and fat have the highest digestibility of any foods on Earth. The better the cut, the more thoroughly the meat is digested and absorbed—to the point where there's almost nothing to egest (opposite of ingest). "For high-quality beef, pork, chicken, or fish, digestibility is about ninety percent," says George Fahey, professor of animal and nutritional sciences at the University of Illinois at Urbana-Champaign. Fats are around 94 percent digestible. A 10-ounce sirloin steak generates but a single ounce of, as they say in George Fahey's lab, egesta.* Best of all: the egg. "Few foods," writes Franz J. Ingelfinger, a panelist at the 1964 Conference on Nutrition in Space and Related Waste Problems, "are digested and assimilated as completely as a hard-boiled egg." That's one reason NASA's traditional launch day breakfast is steak and eggs.† An astronaut may be lying on his back, fully suited, for eight hours or more. You do not want to be eating Fiber One the morning before liftoff. (The Soviet space agency did not traditionally give cosmonauts steak and eggs before launch; it gave them a one-liter enema.)

Fahey, my residue expert, consults for the pet food industry. These are the animal sciences people that NASA should have been working with, not the Air Force vets. The top two concerns of the pet food manufacturer? Palatability and "fecal characteristics": a

---

* *Egesta* is my new favorite euphemism for "feces," and an even better toilet brand name than Ejecto. Certainly better than Toto. Who names a toilet after a lapdog? Unless it's Shit-Tzu. I'd buy a Shit-Tzu toilet.

† Could astronauts live on steak and eggs? Bad idea. Setting aside cholesterol issues, you'd be missing most vitamins. Fahey pointed out that even wild dogs don't live on protein alone. "When they kill prey, they eat a smorgasbord of things." It is a different smorgasbord than the one at Sweden House. "They'll eat the stomach contents first usually." Since the prey is usually some grazing animal, this is their side of vegetables.

clean bowl and a clean living room carpet. First and foremost, dog owners want to feed their pet something it appears to like. I like to think that is NASA's goal as well. "And the number-two concern," said Fahey, setting up a joke he had not intended to make, "is stool consistency. We like to have a fecal material that is hard enough to be picked up and disposed of easily. Not some big mass of runny stuff." Ditto the Gemini and Apollo astronauts.

Pet-food makers also share the early space food scientists' goal of low "defecation discharge frequency." A dog in a high-rise apartment has but two discharge opportunities: once in the morning before its owner leaves for work, and again in the evening. "They have to be able to hold it eight hours," says Fahey. Just like the astronaut on the launch pad. Or the astronaut hoping to put as much time as possible between encounters with the fecal bag.

The other way to lower discharge frequency might be to choose a mellow breed of astronaut. Hyperactive dogs have fast metabolisms; food passes through quickly, so it doesn't have a chance to be completely digested. Hunting dogs, high-strung by nature, tend to have runny stools. And because they're programmed to bound off after prey at any given moment, they wolf their food (no doubt the origin of the verb). This compounds the problem. The less you chew your food, the more of it passes through undigested.

What would Fahey have fed the early astronauts? As a starch, he recommended rice, because it's the lowest-residue of all the carbohydrates. (This is why Purina makes Lamb & Rice, not Lamb & Fingerling Potatoes.) Fresh fruits and vegetables he'd skip, as they create a high-volume, high-frequency stool situation. On the other hand, if you feed someone highly processed foods with no residue, no fiber at all, they'll be constipated. Which, depending on the length of the flight, could be ideal:

"Under current conditions," wrote Franz Ingelfinger, "with the emphasis on short-term flights, I am sure that the most practical solution to the waste-disposal problem has been a constipated astronaut."

TWELVE YEARS AFTER the corned-beef-sandwich incident, astronaut John Young yet again embarrassed his employer in the national news media. Young, along with Apollo 16 crewmate Charlie Duke, was sitting in the Lunar Module Orion after a day out and about collecting rocks. During a radio debriefing with Mission Control, out of the blue, Young declares, "I got the farts again. I got 'em again, Charlie. I don't know what the hell gives them to me. . . . I think it's acid in the stomach." Following Apollo 15, in which low potassium levels were blamed for the heart arrhythmias of the crew, NASA had put potassium-laced orange, grapefruit, and other citrus drinks on the menu.

Young kept going. It's all there in the mission transcript. "I mean, I haven't eaten this much citrus fruit in 20 years. And I'll tell you one thing, in another 12 fucking days, I ain't never eating any more. And if they offer to serve me potassium with my breakfast, I'm going to throw up. I like an occasional orange, I really do. But I'll be damned if I'm going to be buried in oranges." Moments later, Mission Control comes on the line and provides Young with yet more fodder for indigestion.

CAP COM [capsule communicator]: Orion, Houston.

YOUNG: Yes, sir.

CAP COM: Okay, you [have] a hot mike.

YOUNG: Oh. How long have we had that?

CAP COM: It's been on through the debriefing.

This time, it wasn't Congress that got riled. The day after Young's comments hit the press, the governor of Florida issued a statement in defense of his state's key crop, which Charlie Duke paraphrases in his memoir: "It is not our orange juice that is causing the trouble. It's an artificial substitute that doesn't come from Florida."

In fact, it was the potassium, not the orange juice. The "coefficient of flatulence" for orange juice—to use the terminology of USDA flatus researcher Edwin Murphy, another panelist at the 1964 Conference on Nutrition in Space and Related Waste Problems—is low.

Murphy reported on research he had done using an "experimental bean meal" fed to volunteers who had been rigged, via a rectal catheter, to outgas into a measurement device. He was interested in individual differences—not just in the overall volume of flatus but in the differing percentages of constituent gases. Owing to differences in intestinal bacteria, half the population produces no methane. This makes them attractive as astronauts, not because methane stinks (it's odorless), but because it's highly flammable. (Methane is what utility companies sell, under the rubric "natural gas.")*

Murphy had a unique suggestion for the NASA astronaut selection committee: "The astronaut may be selected from that part of our population producing little or no methane or hydrogen"—hydrogen is also explosive—"and a very low level of hydrogen sulfide or other malodorous trace flatus constituents not yet identified. . . . Further, since some individual astronauts may vary in the degree of flatulent reaction to a given weight of food, individuals can be chosen who demonstrate a high resistance to intestinal upset and flatus formation."

---

* If you're among the 50 percent of the population who produce methane, you can play human pilot light. Your friends can hold a match to your gas and watch it ignite and burn blue.

In his work, Murphy had encountered one such ideal astronaut candidate. "Of special interest for further research was the subject who produced essentially no flatus on 100 grams dry weight of beans." As opposed to the average gut, which will, during the peak flatulence period (five to six hours post–bean consumption) pass anywhere from one to almost three cups of flatus per hour. At the high end of the range, that's about two Coke cans full of fart. In a small space where you can't open the window.

As an alternative to recruiting the constitutionally nonflatulent, NASA could create non-"producers" by sterilizing their digestive tract. Murphy had fed the notorious bean meal to a subject who was taking an antibacterial drug and found that the man expelled 50 percent less gas. The saner approach, and the one NASA actually took, was to simply avoid your high coefficient-of-flatus foods. Up through Apollo, beans, cabbage,* Brussels sprouts, and broccoli were blacklisted. "Beans were not used until Shuttle," states Charles Bourland.

There are those who welcomed their arrival, and not just because they're tasty. The zero-gravity fart has been a popular orbital pursuit, particularly on all-male flights. One hears tell of astronauts using intestinal gas like rocket propellant to "launch themselves across the middeck," as astronaut Roger Crouch put it. He had heard the claims and was dubious. "The mass and velocity of the expelled gas," he told me in an email that has forevermore endeared him to me, "is very small compared to the mass of the human body." Thus it was unlikely that it could accelerate a

---

* Cabbage resurfaced in the form of kimchi—fermented spiced cabbage—on board the International Space Station when Korea's first astronaut visited. Space kimchi developer Lee Ju-woon works at the Korean Atomic Energy Research Institute, where scientists are developing ways to harness energy from intestinal kimchi fission. No, they aren't. But they should be.

180-pound astronaut. Crouch pointed out that an exhaled breath doesn't propel an astronaut in any direction, and the lungs hold about six liters of air—versus the fart, which, as we learned from Dr. Murphy, holds at most three soda cans' worth.

Or the average person's, anyway. "My genes have blessed me with an extraordinary ability to expel some of the byproducts of digestion," wrote Crouch. "So given that, I thought that it should be tested. In what I thought was a real voluminous and rapidly expelled purge, I failed to move noticeably." Crouch surmised that his experiment may have been compromised by the "action/reaction of the gas passing through the pants." Disappointingly, both his flights were mixed-gender, so Crouch was disinclined to "strip down naked" and try it again. He was heading to Cape Canaveral and promised to ask around for some other astronauts' input, but so far no one is, as they say, spilling the beans.

ASTRONAUT FOOD IN recent decades has grown kinder and more normal. Meals no longer have to be compressed or dehydrated, as there's plenty of storage room on the International Space Station. Entrées are sealed in plastic pouches, thermostabilized, and then reheated in a small unit that resembles a briefcase. With the 2010 publication of Charles Bourland's incomparable *Astronaut's Cookbook,* it is now possible to whip up eighty-five high-fidelity shuttle-era entrées and sides in your own kitchen, should your own kitchen happen to contain "National 150 filling starch aid from National Starch and Chemical Company" and "caramelized garlic base #99-404 from Eatem Foods."

For a Mars mission, however, things may get strange all over again.

16

EATING YOUR PANTS

Is Mars Worth It?

I will tell you sincerely and without exaggeration that the best part of lunch today at the NASA Ames cafeteria is the urine. It is clear and sweet, though not in the way mountain streams are said to be clear and sweet. More in the way of Karo syrup. The urine has been desalinated by osmotic pressure. Basically it swapped molecules with a concentrated sugar solution. Urine is a salty substance (though less so than the NASA Ames chili), and if you were to drink it in an effort to rehydrate yourself, it would have the opposite effect. But once the salt is taken care of and the distasteful organic molecules have been trapped in an activated charcoal filter, urine is a restorative and surprisingly drinkable lunchtime beverage. I was about to use the word *unobjectionable*, but that's not accurate. People object. They object a lot.

"It makes me sick to have urine in the refrigerator," said my husband Ed. I had finished running yesterday's output through the charcoal and the osmosis bag, and had placed it, in a glass bottle, on the door of the fridge pending lunch down in Mountain

View. I replied that everything objectionable had been filtered out, and that astronauts don't mind drinking treated urine. Ed made a flaring motion with his nostrils and said that circumstances would have to be "postapocalyptic" for him to consider it.

My lunch date at Ames is Sherwin Gormly, a waste-water engineer who helped design the rig to recycle urine on the International Space Station. He has been referred to in the press as "the urine king." This doesn't bother him. What bothered him was being known, briefly, as the guy who said that the moon might be a good place to store weapons-grade plutonium out of reach of megalomaniacal despots. It wasn't a serious suggestion; just Gormly idly speculating. That's what they do down at Ames. In case you didn't pick this up from Norbert Kraft, the NASA of Ames is a different critter from the NASA of Johnson Space Center. "We're a think tank here at Ames," says Gormly. "We're kinda the wingnuts." Gormly is dressed in cargo pants and a lavender Henley shirt. There's nothing especially radical about cargo pants and lavender shirts, but in four trips to Johnson Space Center, I never saw either. Gormly is fit and tan. You'd have to inspect him closely to guess his age correctly; some gray creeping into the blond crewcut, the eyebrows just starting to sprout crazies.

We're not scheduled to land on Mars until sometime in the 2030s, but it's always at the back of the collective NASA mind. The things dreamt up for a lunar base these past five years were dreamt with an eye on Mars. Much of the most innovative stuff comes out of Ames. Not that it will all fly. "Nothing we do," says Gormly, "becomes a space reality until it goes through some filters downstream." You probably want to run anything Sherwin Gormly gives you through some filters.

Landing a spacecraft on Mars is yesterday's challenge. Space agencies have been blasting landers to Mars for three decades. (Remember, once a craft reaches space, there's no air drag to slow it down; it keeps traveling through the vacuum of space without needing more rocket power, aside from small course corrections. Space ships basically coast to Mars. The fuel they'd need is for landing and for the return blast back.) Rockets powerful enough to accelerate an 800-pound lander to Mars are a whole other animal from a rocket that can do so while carrying five or six humans and two-plus years' worth of supplies.

Back in the sixties, when aerospace scientists assumed that the follow-up to a moon landing would be a manned Mars mission, some fantastical Ames-style creativity was afoot. An obvious alternative to launching 8,000 pounds of food is to grow it—or some of it—on board in greenhouses. But in the early sixties, meat ruled the dinner plate. The space nutritionists, for a brief and wondrous moment, turned their minds to the possibility of zero-gravity ranching. "What type of animal should be taken along to Mars or Venus?" asked animal husbandry professor Max Kleiber at the 1964 Conference on Nutrition in Space and Related Waste Problems. Kleiber held an accommodating view of animal husbandry; he included rats and mice in his calculations along with cattle and sheep. He left the unpretty logistics of zero-gravity slaughter and manure management to others, for Kleiber was a metabolism man. He simply wished to know: Which beast provides the greatest number of calories for the lowest launch weight and feed consumption? To serve beef to two or three Mars astronauts, "a steer of 500-kilogram body weight has to be hauled into space." Whereas the same number of calories could be derived from just 42 kilograms of mice (about 1,700 of them). "The astronauts," stated the paper's conclusion, "should eat mouse stew instead of beef steaks."

Present at the same conference was D. L. Worf, of the Martin Marietta Company (before Lockheed got there). Worf was big on thinking outside the box, and then eating it. "Food may be processed by many of the same techniques that are used to fabricate structures and shapes from plastic." Worf did not limit this thinking to food containers but included spacecraft structures normally jettisoned or left behind when preparing to return home. In other words, instead of abandoning the Lunar Module on the moon, the Apollo 11 crew could have broken off pieces to take along and eat on the way home. Thereby needing to carry less food in the first place. Worf envisioned a return-trip menu that included Fuel Tank, Rocket Motor, and Instrument Casing. Leave room for dessert! "Transparent sugar castings as a substitute for windows" also made Worf's idea list.

You wouldn't complain about a breakfast of Worf's egg-albumin office paper if you'd sampled Dr. Carl Clark's paper cuisine. Clark, a Navy biochemist, was quoted in a 1958 *Time* article on long-duration spaceflight, recommending that astronauts add shredded paper—the ordinary wood pulp variety—as a "thickener" to a main course of vitamin- and mineral-enriched sugar water. Whether Clark viewed the shredded paper as an aid to palatability, regularity, or document security, I can't say.

"If the imagination is allowed to wander"—and with D.L. Worf it surely should be—astronauts could also eat their dirty clothes. Worf estimated that "a space crew of four men will, for a 90-day flight regime, dispose of approximately 120 pounds of clothing, if laundry facilities are not available." (Thanks in large part to Sherwin Gormly, they now are.) For a three-year Mars mission, that's 1,440 pounds of dirty wash/victuals. Worf reported that several companies were already spinning textiles from soybeans and milk proteins and that the U.S. Department of Agriculture has "prepared [textile] fibers from egg whites and chicken feathers that would be highly acceptable as food under the controlled environment of a spacecraft." Meaning, I think, that a man who is willing to dine on used clothing is a man unlikely to balk at chicken feathers.

But why go to the added expense of shopping at USDA experimental research stations? "Keratin protein fibers such as wool and silk," muses Worf, "could be converted to food by partial hydrolysis. . . ."

Onboard hydrolysis is the point where astronauts start to get uncomfortable. Hydrolysis is a process by which proteins, edible if not necessarily palatable, are broken down into still edible but typically less palatable constituents. Vegetable protein, for instance, can be hydrolyzed to make MSG. Pretty much any amino acid arrangement can be hydrolyzed, including those of the recyclable that dares not speak its name. A four-person crew will, over the course of three years, generate somewhere in the neighborhood of a thousand pounds of feces. In the ominous words of sixties space nutritionist Emil Mrak, "The possibility of reuse must be considered."

Sometime in the early 1990s, University of Arizona microbiologist Chuck Gerba was invited to a Martian strategy workshop whose topics included solid-waste management. Gerba told me that he recalls one of the chemists saying, "Shoot, what we could do is hydrolyze the stuff back to carbon and make patties out of it." Whereupon the astronauts in attendance went, "We are *not* eating shit burgers on the way back."

Moralewise, this brand of extreme recycling is ill advised. The current Mars thinking is to deposit caches of food ahead of time, using unmanned landers. (The strategy of leaving caches on Mars came up during an interview with some Russian cosmonauts. My interpreter Lena paused and said, "Mary, what did you say about kasha on Mars?")

A better way to recycle astronautical by-product would be to seal it into plastic tiles and use it as shielding against cosmic radiation. Hydrocarbons are good for this. Metal spacecraft hulls are not; radiation particles break down into secondary particles as they pass through. These fragmented bits can be more dangerous than the intact primary particles. So what if you'd be, as Gerba crowed, "flying in shit!" Beats leukemia.

• • •

GORMLY AND I have been talking about psychological barriers to progress. As it turns out, we're not the only Californians drinking treated urine this afternoon. (In solidarity, Gormly treated a batch of his own.) The citizens of Yellow, I mean Orange, County are drinking it right along with us. The difference, says Gormly, is that Orange County pumps theirs into the ground for a while before they call it drinking water again. "There is absolutely no technical justification for what they're doing. It's psychosocial and political," he says. People are not ready for "toilet to tap."

Even here at Ames. As Gormly stood in line to pay for his sandwich, the man ahead of us asked what was in the bottle. "It's treated urine," said Gormly, straight-faced but obviously enjoying himself. The man glanced at Gormly, looking for something that might confirm the hope that Gormly had made a joke. "No, it's not," he decided and walked away.

The cashier was going to be tougher. "What did you say was in the bottle?" She looked like she might be wanting to call security.

This time Gormly said, "Life support experiment." Confronted with science, the woman backed down.

One of the things I love about manned space exploration is that it forces people to unlace certain notions of what is and isn't acceptable. And possible. It's amazing what sometimes gets accomplished via an initially jarring but ultimately harmless shift in thinking. Is cutting the organs out of a dead man and stitching them into someone else barbaric and disrespectful, or is it a straightforward operation that saves multiple lives? Does crapping into a Baggie while sitting 6 inches away from your crewmate represent a collapse of human dignity or a unique and comic form of intimacy? The latter, by Jim Lovell's reckoning. "You get to know each other so well you don't even bother turning away." Your wife and kids have seen you on the toilet. So Frank Borman sees you. Who cares? Worth it for the prize at the bottom of the box.

When someone tells a crew of astronauts they're going to have to drink treated sweat and urine—not only their own, but that of their crewmates and, who knows, the 1,700 mice in the pantry, they shrug and say, "No biggie." Maybe astronauts aren't just expensive action figures. Maybe they're the poster boys and girls for the new environmental paradigm. As Gormly says, "Sustainability engineering and human spaceflight engineering are just different sides of the same technology."

The tougher question is not "Is Mars possible?" but "Is Mars worth it?" An outside estimate of the cost of a manned mission to Mars is roughly the cost of the Iraq war to date: $500 billion. Is it similarly hard to justify? What good will come of sending humans to Mars, especially when robotic landers can do a lot of the science just as well, if not as fast? I could parrot the NASA Public Affairs Office and spit out a long list of products and technologies*

---

* If it's cordless, fireproof, lightweight and strong, miniaturized, or automated, chances are good NASA has had a hand in the technology. We are talking trash compactors, bulletproof vests, high-speed wireless data transfer, implantable heart monitors, cordless power tools, artificial limbs, dustbusters, sports bras, solar panels, invisible braces, computerized insulin pumps, firefighters' masks. Every now and then, earthbound applications head off in an unexpected direction: Digital lunar image analyzers allow Estée Lauder to quantify "subtleties otherwise undetectable" in the skin of women using their products, providing a basis for ludicrous wrinkle-erasing claims. Miniature electronic Apollo heat pumps spawned the Robotic Sow. "At feeding time a heat lamp simulating a sow's body warmth is automatically turned on, and the machine emits rhythmic grunts like a mother pig summoning her piglets. As piglets scamper to their mechanical mother, a panel across the front opens to expose the row of nipples," wrote an unnamed NASAfacts scribe, surely eliciting grunts from superiors in the NASA Public Affairs Office.

spawned by aerospace innovations over the decades. Instead, I defer to the sentiments of Benjamin Franklin. Upon the occasion of history's first manned flights—in the 1780s, aboard the Montgolfier brothers' hot-air balloons—someone asked Franklin what use he saw in such frivolity. "What use," he replied, "is a newborn baby?"

It might not be that hard to raise the funds. If the nations involved were to approach their respective entertainment conglomerates, an impressive hunk of funding could be raised. The more you read about Mars missions, the more you realize it's the ultimate reality TV.

I was at a party the day the Phoenix robotic lander touched down on Mars. I asked the party's host, Chris, if he had a computer I could use to watch the NASA TV coverage. At first it was just Chris and I watching. By the time Phoenix had plowed intact through the Martian atmosphere and was about to release its parachute for the descent, half the party was upstairs crowded around Chris's computer. We weren't even watching Phoenix. The images hadn't yet arrived. (It takes about twenty minutes for signals to travel between Mars and Earth.) The camera was trained on Mission Control at the Jet Propulsion Laboratory. It was standing-room with engineers and managers, people who'd spent years working on heat shields and parachute systems and thrusters, all of which, in this final hour, could fail in a hundred different ways, each of those failures having been planned for with backup hardware and contingency software. One man stared at his computer with the fingers of both hands crossed. The touchdown signal arrived, and everyone was up on their feet making noise. Engineers bear-hugged each other so enthusiastically that they knocked their glasses crooked. Someone began passing out cigars. We all yelled too and some of us got a little choked up. It was inspiring, what these men and women had done. They flew a delicate scientific instrument more than 400 million miles to Mars and set it down as gently as a baby, exactly where they wanted it.

We live in a culture in which, more and more, people live through simulations. We travel via satellite technology, we socialize on computers. You can tour the Sea of Tranquility on Google Moon and visit the Taj Mahal via Street View. Anime fans in Japan have been petitioning the government for the right to legally marry a two-dimensional character. Fundraising has begun on a $1.6 billion resort in the rim of a simulated Martian crater in the desert outside Las Vegas. (They can't simulate Martian gravity, but the boots of the spacesuits will be "a little more bouncy.") No one goes out to play anymore. Simulation is becoming reality.

But it isn't anything like reality. Ask an M.D. who spent a year dissecting a human form tendon by gland by nerve, whether learning anatomy on a computer simulation would be comparable. Ask an astronaut whether taking part in a space simulation is anything like being in space. What's different? Sweat, risk, uncertainty, inconvenience. But also, awe. Pride. Something ineffably splendid and stirring. One day at Johnson Space Center, I visited Mike Zolensky, the curator of cosmic dust and one of the caretakers of NASA's meteorite collection. Every now and then, a piece of asteroid slams into Mars hard enough that the impact hurls small chunks of the Martian surface way out into space, where they continue to travel until they are snagged by some other planet's gravitational pull. Occasionally that planet is Earth. Zolensky opened a case and lifted out a Martian meteorite as heavy as a bowling ball and handed it to me. I stood there taking in its hardness and heft, its *realness,* making an expression that I'm sure I'd never before had call to make. The meteorite wasn't beautiful or exotic-looking. Give me a chunk of asphalt and some shoe polish and I can make you a simulated Mars meteorite. What I can't possibly simulate

ACKNOWLEDGMENTS

The first time I visited Johnson Space Center, a sign near the door of the public affairs building said, HARD HAT REQUIRED. And it kind of was. A lot of No's got lobbed my way. Space agencies keep a firm grip on their public image, and it's less troublesome for employees and contractors to say no to someone like me than to take their chances and see what I write. Happily there are people involved in the human side of space exploration who see value in unconventional coverage (or are just plain too nice to say no). For their candor and wit—and the generosity with which they shared their time and know-how—super-galactic thanks to John Bolte, Charles Bourland, James Broyan, John Charles, Tom Chase, Jon Clark, Sherwin Gormly, Ralph Harvey, Norbert Kraft, Rene Martinez, Joe Neigut, Don Rethke, and Scott Weinstein; astronauts Roger Crouch, Jim Lovell, Lee Morin, Mike Mullane, Andy Thomas, and Peggy Whitson; and in Russia, cosmonauts Sergei Krikalyov, Alexandr Laveikin, Yuri Romanenko, and Boris Volynov.

I have no background in space or aeromedical matters. Many of the people I spoke to were not so much sources as unpaid tutors. I am talking about Dennis Carter, Pat Cowings, Seth Donahue, George Fahey, Brian Glass, Dustin Gohmert, Sean Hayes, Toby Hayes, Natsuhiko Inoue, Nick Kanas, Tom Lang, Pascal Lee, Jim Leyden, Marcelo Vazquez, April Ronca, Charles Oman, Brett Ringger, Shoichi Tachibana, Art Thompson, Nick Wilkinson, and

Mike Zolensky. All spent more time with me than they had to spare, and for this I am truly grateful.

Terry Sunday's tremendous expertise and thoughtful, thorough review of the manuscript and Linda Wang's knowledge of congressional archives were indispensable. For their insights into things that happened long ago, I am grateful to Bill Britz, Earl Cline, Jerry Fineg, Dan Fulgham, Wayne Mattson, Joe McMann, May O'Hara, Rudy Purificato, and Michael Smith. Pam Baskins, Simone Garneau, Jenny Gaultier, Amy Ross, Andy Turnage, and Violet Blue provided valuable contacts and assistance, and I thank them too.

Though the public affairs people could not always help in the ways I naively wanted them to, they were extremely knowledgeable and professional. Aaisha Ali, Gayle Frere, James Hartsfield, and Lynnette Madison of the Johnson Space Center were especially attentive, as was Kathryn Major of the National Space Biomedical Research Institute and Trish Medalen at Red Bull. Kumiko Tanabe of the Japan Aerospace Exploration Agency worked miracles on my behalf. I'd also like to acknowledge the people who put together NASA's oral history and Lunar Surface Journal projects and the oral history program at the New Mexico Museum of Space History, as well as the staff of the Interlibrary Loan department of the San Francisco Public Library. These are incomparable resources.

Lena Yakovlena, Sayuri Kanamori, and Manami Tamaoki were not only brilliant interpreters but unbeatable travel companions. I am extremely fortunate that Fred Wiemer was available to copy edit both this and my previous book. Thanks to designer Jamie Keenan for another perfect and witty cover; to curator Deirdre O'Dwyer for the hours spent stalking obscure photos and rights; to the fabulous Kristen Engelhardt for spot translations; to the bed-resters for their boundless good humor; to Jeff Greenwald for books, gin, and enthusiasm; and to Dan Menaker for the best line in the book.

As with all my books, any success must be attributed in large part to the collective publishing chops of W. W. Norton. With the help of a dorky rocket metaphor, I would like to single a few people out. My incomparable editor Jill Bialosky deftly steered the manuscript through some needed midcourse corrections, and Rebecca Carlisle, Erin Sinesky Lovett, and Steve Colca expertly managed launch and trajectory of the finished product.

My husband Ed Rachles and my agent Jay Mandel gracefully defused the angst and whinging pessimism that are an inevitable part of all my ventures. I don't think I could do what I do without the support of these two excellent people.

TIME LINE

1949	Rhesus monkey Albert II becomes first creature to experience zero gravity on board a rocket.
1950–1958	Air Force flies planes in parabolas to mimic zero G and study its effects on chimps, cats, humans.
Nov. 1957	Soviet dog Laika orbits Earth, dies in space.
Aug. 1960	Soviet dogs Belka and Strelka are first to return alive from orbit.

Mercury Space Program Era 1961–1963

Jan. 31, 1961	Astrochimp Ham survives a suborbital flight in a Mercury space capsule.
April 12, 1961	Yuri Gagarin becomes the first human in space, and first human to orbit Earth.
May 5, 1961	Alan Shepard becomes first American in space.
Nov. 29, 1961	Astrochimp Enos orbits Earth.
Feb. 20, 1962	John Glenn becomes the first American to orbit Earth.

Gemini Space Flights 1965–1966	
1965–1966	Air Force tests Gemini diets and "restricted bathing" regimens in space cabin simulators.
Mar. 18, 1965	Alexei Leonov becomes first astronaut to spacewalk outside spacecraft.
Mar. 23, 1965	Gemini III "corned beef sandwich incident."
June 3, 1965	Gemini IV: Ed White becomes NASA's first spacewalker.
Dec. 4–18, 1965	Gemini VII: two men, two weeks, no bath.

Apollo Lunar Missions 1968–1972	
Mar. 3–13, 1969	Apollo 9: Rusty Schweickart battles space motion sickness.
July 20, 1969	Apollo 11: first humans set foot on the moon.
Dec. 7–9, 1972	Apollo 17: first scientist in space.

Orbiting Space Station (and Space Shuttle) Era 1973–2015	
1973–1979	Skylab U.S. space station missions; space showers prove untenable.
1971–1982	Salyut Soviet space station missions.
Jan. 1978	First U.S. female astronaut candidate.
April 12, 1981	First Space Shuttle launch.
Jan. 28, 1986	Space Shuttle Challenger disaster.
1986–2001	Mir.
Nov. 2000	First International Space Station mission.
Feb. 1, 2003	Space Shuttle Columbia disaster.

BIBLIOGRAPHY

COUNTDOWN

Gagarin, Yuri. *Road to the Stars.* Moscow: Foreign Languages Publishing House, 1962. P. 170.

Gemini VII Voice Communications: Air to Ground, Ground to Air, and On-Board Transcription. Vol. 1, p. 239. NASA History Portal: http://www.jsc.nasa.gov/history/mission_trans/gemini7.htm.

Platoff, Anne M. "Where No Flag Has Gone Before: Political and Technical Aspects of Placing a Flag on the Moon." NASA Contractor Report 188251. August 1993.

1 HE'S SMART BUT HIS BIRDS ARE SLOPPY

Cernan, Eugene, and Don Davis. *The Last Man on the Moon.* New York: St. Martin's Press, 1999. Pp. 308–310.

Mullane, Mike. *Riding Rockets: The Outrageous Tales of a Space Shuttle Astronaut.* New York: Scribner, 2006. Pp. 191, 297.

Pesavento, Peter. "From Aelita to the International Space Station: The Psychological Effects of Isolation on Earth and in Space." *Quest: The History of Spaceflight Quarterly* 8 (2): 4–23 (2000).

Santy, Patricia. *Choosing the Right Stuff: The Psychological Selection of Astronauts and Cosmonauts.* Westport, Conn.: Praeger, 1994.

Zimmerman, Robert. *Leaving Earth: Space Stations, Rival Superpowers, and the Quest for Interplanetary Travel.* Washington, D.C.: Joseph Henry Press, 2006.

2 LIFE IN A BOX

Ackmann, Martha. *The Mercury 13: The True Story of Thirteen Women and the Dream of Space Flight.* New York: Random House, 2004.

"Airman Still Okay in Mock Trip to Moon." *Hayward Daily Review,* February 10, 1958.

Burnazyan, A. J., et al. "Year-Long Medico-Engineering Experiment in a Partially Closed Ecological System." *Aerospace Medicine,* October 1969, pp. 1087–1093.

Collins, Michael. *Liftoff: The Story of America's Adventure in Space.* New York: Grove Press, 1988. P. 262.

Gemini VII Composite Air-to-Ground and Onboard Voice Tape Transcription. Vol. 2, pp. 461, 500. NASA History Portal: http://www.jsc.nasa.gov/history/mission_trans/gemini7.htm.

Kanas, Nick, and Dietrich Manzey. *Space Psychology and Psychiatry*, 2nd ed. El Segundo, Calif.: Microcosm Press, 2008.

Malik, Tariq. "Report: Russia's Mock Mars Mission to Cost $15 Million." SPACE.com, posted January 7, 2008. http://www.space.com/news/080107-russia-esa-mockmars-cost.html.

Nowak, Lisa. Orange County Charging Affidavit. Reproduced on the Smoking Gun Web site.

Pesavento, Peter. "From Aelita to the International Space Station: The Psychological Effects of Isolation on Earth and in Space." *Quest: The History of Spaceflight Quarterly* 8 (2): 4–23 (2000).

Stuster, Jack. "Space Station Habitability Recommendations Based on a Systematic Comparative Analysis of Analogous Conditions." NASA Contractor Report 3943 (NASA-CR 3943).

Zimmerman, Robert. *Leaving Earth: Space Stations, Rival Superpowers, and the Quest for Interplanetary Travel*. Washington, D.C.: Joseph Henry Press, 2006.

3 STAR CRAZY

Cernan, Eugene, and Don Davis. *The Last Man on the Moon*. New York: St. Martin's Press, 1999. Pp. 132–144.

Clark, Brant, and Ashton Graybiel. "The Break-Off Phenomenon: A Feeling of Separation from the Earth Experienced by Pilots at High Altitude." *Aviation Medicine* 28 (2): 121–126 (1957).

Gemini IV Composite Air-to-Ground and Onboard Voice Tape Transcription. Pp. 43–57. NASA History Portal: http://www.jsc.nasa.gov/history/mission_trans/gemini4.htm.

Gussow, Zachary. "A Preliminary Report of Kayak-Angst Among the Eskimo of West Greenland: A Study in Sensory Deprivation." *International Journal of Social Psychiatry* 9: 18–26 (1963).

Kanas, Nick, and Dietrich Manzey. *Space Psychology and Psychiatry*, 2nd ed. El Segundo, Calif.: Microcosm Press, 2008.

Linenger, Jerry M. *Off the Planet: Surviving Five Perilous Months Aboard the Space Station Mir.* New York: McGraw Hill, 2000. P. 147.

Oman, Charles. "Spatial Orientation and Navigation in Microgravity." In *Spatial Processing in Navigation, Imagery, and Perception.* Edited by Fred Mast and Lutz Jancke. New York: Springer, 2007.

Portree, David S. F., and Robert C. Trevino. "Walking to Olympus: An EVA Chronology" (Monographs in Aerospace History Series #7). Washington, D.C.: NASA History Office, 1997.

Shayler, David J. *Disasters and Accidents in Manned Spaceflight.* New York: Springer-Praxis, 2000.

Simons, David G., with Dan A. Schanche. *Man High.* Garden City, N.Y.: Doubleday, 1960.

Zimmerman, Robert. *Leaving Earth: Space Stations, Rival Superpowers, and the Quest for Interplanetary Travel.* Washington, D.C.: Joseph Henry Press, 2006. P. 108.

4 YOU GO FIRST

Burgess, Colin, and Chris Dubbs. *Animals in Space: From Research Rockets to the Space Shuttle.* Chichester, U.K.: Springer-Praxis, 2007.

Debruicker, John. "Anti-Gravity Stone Has a Strange Story and an Even Stranger History." *Colby* [College] *Echo*, March 9, 2006.

Gillespie, Charles. *The Montgolfier Brothers and the Invention of Aviation.* Princeton, N.J.: Princeton University Press, 1983.

Girifalco, Louis A. *The Universal Force: Gravity—Creator of Worlds.* Oxford: Oxford University Press, 2007.

Helmore, Edward. "Timothy Leary's Final Trip: Boldly Going into Orbit." *The Independent* (UK), April 21, 1997, online ed.

Kittinger, Joe. Space Center Oral History Program. Interviewed by Wayne O. Mattson and George M. House at the New Mexico Museum of Space History, Alamogordo, New Mexico, November 2000.

Simons, David. Space Center Oral History Program. Interviewed by George P. Kennedy at the International Space Hall of Fame, Alamogordo, New Mexico, September 1987.

von Beckh, H. J. A. "Experiments with Animals and Human Subjects Under Sub- and Zero-Gravity Conditions During the Dive and Parabolic Flight." *Aviation Medicine*, June 1954. Pp. 235–241.

Ward, Julian, Willard Hawkins, and Herbert Stallings. "Physiologic Response to Subgravity: Initiation of Micturition." *Aerospace Medicine*, August 1959.

———. "Physiologic Response to Subgravity: Mechanics of Nourishment and Deglutition of Solids and Liquids." *Aviation Medicine*, March 1959.

5 UNSTOWED

Ayres, M. L. "Survival After Jet Engine Intake." *Injury* 4: 317–318.

Collins, Michael. *Flying to the Moon: An Astronaut's Story*, 2nd ed. New York: Farrar, Straus & Giroux (Sunburst Book), 1994. Pp. 80–81.

6 THROWING UP AND DOWN

Apollo 9 Onboard Voice Transcription, Vol. 1, Day 2. NASA History Portal: http://www.jsc.nasa.gov/history/mission_trans/apollo9.htm.

Apollo 16 Lunar Surface Journal. http://history.nasa.gov/alsj/a16/a16.html.

Apollo 16 Onboard Voice Transcription, Lunar Module, Day 5. NASA History Portal: http://www.jsc.nasa.gov/history/mission_trans/apollo16.htm.

Brown, Tony. *Hendrix: The Final Days.* London and New York: Omnibus Press, 1997.

Cernan, Eugene, and Don Davis. *The Last Man on the Moon*. New York: St. Martin's Press, 1999. Pp. 120, 190.

Correia, M. J., and F. E. Guedry, Jr. "Modification of Vestibular Responses as a Function of Rate of Rotation About an Earth-Horizontal Axis." *Acta-oto-laryngologica* 62: 297–304.

Gell, C. F. and D. Cranmore. "The Effects of Acceleration on Small Animals Utilizing a Quick-Freeze Technique." *Aviation Medicine*, February 1953, pp. 48–56.

Harsch, Viktor. "Centrifuge 'Therapy' for Psychiatric Patients in Germany in the Early 1800s." *Aviation, Space, and Environmental Medicine* 77(2): 157–160 (2006).

Money, K. E. "Motion Sickness." *Physiological Reviews* 50 (1): 1–35.

Neale, Richard. Letters to the editor in *Lancet*, February 19, 1887, p. 403.

Noble, R. L. "Observations on Various Types of Motion-Causing Vomiting in Animals." *Canadian Journal of Research* 23 (E): 212–219.

Oman, Charles. "Spatial Orientation and Navigation in Microgravity." In *Spatial Processing in Navigation, Imagery, and Perception.* Edited by Fred Mast and Lutz Jancke. New York: Springer, 2007. Pp. 209–233.

Oman, Charles, Byron K. Lichtenberg, and Kenneth E. Money. "Space Motion Sickness Monitoring Equipment: Spacelab 1." Paper reprinted from Conference Proceedings No. 372 (Motion Sickness: Mechanisms, Prediction, Prevention, and Treatment) of the Advisory Group for Aerospace Research and Development.

Rannie, Ian. "The Effect of the Inhalation of Vomitus on the Lungs: Morbid Anatomy." *British Journal of Anaesthesia* 35: 146 (1963).

Reason, J. T, and J. J. Brand. *Motion Sickness.* London and New York: Academic Press, 1975.

Schweickart, Russell L. Oral history. Johnson Space Center Oral History Project. http://www.jsc.nasa.gov/history/oral_histories/oral_histories.htm.

Thurston, Paget. Letter to the editor in *Lancet,* February 12, 1887, p. 350.

Vandenberg, James T., et al. "Large-Diameter Suction Tubing Significantly Improves Evacuation Time of Simulated Vomitus." Paper presented at the annual meeting of the Society of Academic Emergency Medicine, Denver, May 1996.

Wade, Nicholas J., U. Norrsell, and A. Presly. "Cox's Chair: 'A Moral and a Medical Mean in the Treatment of Maniacs.'" *History of Psychiatry* 16 (1): 73–88 (2005).

Wolfe, Robert C., Marcus Reidenberg, and Vicente Dinoso. "Tang and Methadone by Vein." *Annals of Internal Medicine* 76: 830.

7 THE CADAVER IN THE SPACE CAPSULE

AP Wire Service. "Chimps Aid Space Study." *Denton Record-Chronicle,* March 27, 1966.

Brown, William K., Jerry Rothstein, and Peter Foster. "Human Response to Predicted Apollo Landing Impacts in Selected Body Orientations." *Aerospace Medicine*, April 1966, pp. 394–398.

Melvin, John W., et al. "Crash Protection of Stock Car Racing Drivers—Application of Biomechanical Analysis of Indy Car Crash Research." *Stapp Car Crash Journal* 50: 415–428.

Mullane, Mike. *Riding Rockets: The Outrageous Tales of a Space Shuttle Astronaut.* New York: Scribner, 2006. Pp. 330–331.

Swearingen, John J., et al. "Human Voluntary Tolerance to Vertical Impact." *Aerospace Medicine,* December 1960, pp. 989–995.

Walz, Feix H., et al. "Airbag Deployment and Eye Perforation by a Tobacco Pipe." *Journal of Trauma* 38 (4): 498–501 (1995).

8 ONE FURRY STEP FOR MANKIND

Burgess, Colin, and Chris Dubbs. *Animals in Space: From Research Rockets to the Space Shuttle.* Chichester, UK: Springer-Praxis, 2007.

"Chimps Aid Space Study." *Denton Record-Chronicle,* March 27, 1966.

Collins, Michael. *Liftoff: The Story of America's Adventure in Space.* New York: Grove Press, 1989.

Dooling, Dave. "The One-Way Manned Mission to the Moon." *Quest: The History of Spaceflight Quarterly* 8 (4): 4–11.

Glenn, John H., Jr. Oral history. Johnson Space Center Oral History Project. http://www.jsc.nasa.gov/history/oral_histories/oral_histories.htm.

"Kennedy Birthday Party Enlivened by Monkeys." *Albuquerque Tribune,* November 22, 1961.

Siddiqi, Asif, "'There It Is!': An Account of the First Dogs-in-Space Program." *Quest: The History of Spaceflight Quarterly* 5 (3): 38–42.

Swenson, Lloyd, Jr., James M. Grimwood, and Charles C. Alexander. *This New Ocean: A History of Project Mercury.* (NASA SP-4201.) Washington, D.C.: NASA Scientific and Technical Information Division, 1966.

Williams, Harold R. "'Chimp College' Flourishes at Air Base in New Mexico." AP Wire Service: *Hobbs Daily News-Sun,* November 20, 1963.

———"'Chimp College' Graduates Famous All Over Nation." AP Wire Service: *Hobbs Daily News-Sun,* November 21, 1963.

———"First U.S. Flag on Moon May Be Planted by Chimp." AP Wire Service: *Bridgeport Post,* May 18, 1962.

9 NEXT GAS: 200,000 MILES

Apollo 17 Lunar Surface Journal. http://history.nasa.gov/alsj/a17/a17.html.

Cernan, Eugene, and Don Davis. *The Last Man on the Moon.* New York: St. Martin's Press, 1999. Pp. 120, 190.

Coulter, Donna. "Moondust in the Wind." Moondaily.com report, April 14, 2008.

Fox, William L. *Driving to Mars.* Emeryville, Calif: Shoemaker & Hoard, 2006.

Gernhardt, Michael, Andrew Abercromby, and Pascal Lee. "Evaluation of Small Pressurized Rover and Mobile Habitat Concepts of Operations During Simulated Planetary Surface Exploration." Unpublished NASA test plan, 2008.

Kanas, Nick. "High Versus Low Crewmember Autonomy in Space Simulation Environments." Paper presented at the 60th International Astronautical Congress, Daejeon, Korea, October 2009.

10 HOUSTON, WE HAVE A FUNGUS

Berry, Charles A. Oral history. Johnson Space Center Oral History Project. http://www.jsc.nasa.gov/history/oral_histories/oral_histories.htm.

Borman, Frank. Oral history. Johnson Space Center Oral History Project. http://www.jsc.nasa.gov/history/oral_histories/oral_histories.htm.

Cernan, Eugene, and Don Davis. *The Last Man on the Moon.* New York: St. Martin's Press, 1999. P. 95.

Gemini VII Composite Air-to-Ground and Onboard Voice Tape Transcription. Vol. 1–3. NASA History Portal: http://www.jsc.nasa.gov/history/mission_trans/gemini7.htm.

Larson, Elaine. "Hygiene of the Skin: When Is Clean Too Clean?" *Emerging Infectious Diseases* 7 (2) (March/April 2001). http://www.cdc.gov/ncidod/eid/vol7no2/larson.htm.

Lovell, James A., Jr. Oral history. Johnson Space Center Oral History Project. http://www.jsc.nasa.gov/history/oral_histories/oral_histories.htm.

Milunas, Michele C., Ann F. Rhoads, and J. Russell Mason. "Effectiveness of Odour Repellants for Protecting Ornamental Shrubs from Browsing by White-Tailed Deer." *Crop Protection* 13 (5): 393–399.

Popov, I. G., et al. "Investigation of the State of the Human Skin After Prolonged Restriction on Washing." In *Problems of Space Biology,* vol. 7. Edited by V. N. Chernigovsky. Moscow: Nanka Publishing Co., 1969. Pp. 386–392. May 1969.

Slonim, A. R. "Effects of Minimal Personal Hygiene and Related Procedures During Prolonged Confinement." Aerospace Medical Research Laboratories Technical Report AMRL-TR-66-146, October 1966.

Stuster, Jack. "Space Station Habitability Recommendations Based on a Systematic Comparative Analysis of Analogous Conditions." NASA Contractor Report 3943 (NASA-CR 3943).

11 THE HORIZONTAL STUFF

Allen, C., P. Glasziou, and C. Del Mar. "Bed Rest: A Potentially Harmful Treatment Needing More Careful Evaluation." *Lancet* 354: 1229–1234 (1999).

Donahue, Seth W., et al. "Parathyroid Hormone May Maintain Bone Formation in Hibernating Black Bears (*Ursus americanus*) to Prevent Disuse Osteoporosis." *Journal of Experimental Biology* 209: 1630–1638.

Finkelstein, Joel S., et al. "Ethnic Variation in Bone Density in Premenopausal and Early Perimenopausal Women: Effects of Anthropomorphic and Lifestyle Factors." *Journal of Clinical Endocrinology & Metabolism* 87 (7): 3057–3067.

Parker-Pope, Tara. "Drugs to Build Bones May Weaken Them." *New York Times*, Science Times, July 15, 2008.

Shropshire, Courtney. *American Journal of Dermatology* (1912), p. 318.

12 THE THREE-DOLPHIN CLUB

Beier, John C., and Douglas Wartzok. "Mating Behavior of Captive Spotted Seals (*Phoca largha*)." *Animal Behavior* 27: 772–781 (1979).

Levin, R. J. "Effects of Space Travel on Sexuality and the Human Reproductive System." *Journal of the British Interplanetary Society* 47: 378–382 (1989).

Mullane, Mike. *Riding Rockets: The Outrageous Tales of a Space Shuttle Astronaut.* New York: Scribner, 2006. Pp. 176–177.

Pesavento, Peter. "From Aelita to the International Space Station: The Psychological Effects of Isolation on Earth and in Space." *Quest: The History of Spaceflight Quarterly* 8 (2): 4–23 (2000).

Rambaut, Paul, C., et al. "Some Flow Properties of Food in Null Gravity." *Food Technology*, January 1972.

Ronca, April, and Jeffrey R. Alberts. "Physiology of a Microgravity Environment: Selected Contribution—Effects of Spaceflight During Pregnancy on Labor and Birth at 1 G." *Journal of Applied Physiology* 89: 849–854 (2000).

Stine, Harry G. *Living in Space.* New York: M. Evans, 1997.

Snopes.com. "Claim: NASA Shuttle Astronauts Conducted Sex Experiments in Space." http://www.snopes.com/risque/tattled/shuttle.asp.

13 WITHERING HEIGHTS

"Changes in Intracranial Volume on Ascent to High Altitudes and Descent as in Diving." Staff Meetings of the Mayo Clinic, April 2, 1941. Pp. 220–224.

Haber, Fritz. "Bailout at Very High Altitudes." *Aviation Medicine*, August 1952. Pp. 322–329.

Kornfield, Alfred T., and J. R. Poppen. "High Velocity Wind Blast on Personnel and Equipment." *Aviation Medicine*, February 1949. Pp. 24–38.

McAndrew, James. *The Roswell Report: Case Closed*. Washington, D.C.: U.S. Government Printing Office, 1997.

NASA. Columbia Crew Survival Investigation Report. NASA/SP-2008-565. http://www.nasa.gov/pdf/298870main_SP-2008-565.pdf.

Ryan, Craig. *The Pre-Astronauts: Manned Ballooning on the Threshold of Space*. Annapolis, Md.: Naval Institute Press, 1995.

Stapp, John P. "Human Tolerance Factors in Supersonic Escape." *Aviation Medicine,* February 1957. Pp. 77–90.

14 SEPARATION ANXIETY

Apollo 10 Onboard Voice Transcription—Command Module, Day 6, pp. 364–365, 414–420. http://www.jsc.nasa.gov/history/mission_trans/apollo10.htm.

Broyan, James Lee, Jr. "Waste Collector System Technology Comparisons for Constellation Applications." SAE Technical Paper 2007-01-3227.

Wignarajah, Kanapathipillai, and Eric Litwiller. "Simulated Human Feces for Testing Human Waste Processing Technologies in Space Systems." SAE Technical Paper 2006-01-2180, presented at the 36th International Conference on Environmental Systems, Norfolk, Va. July 17–20, 2006.

15 DISCOMFORT FOOD

"A Guideline of Performing *Ibadah* at the International Space Station (ISS)." The Islamic Workspace blog: http://makkah.wordpress.com.

Apollo 16 Mission Commentary. NASA History Portal: http://www.jsc.nasa.gov/history/mission_trans/apollo16.htm.

Bourland, Charles T. Oral history. Johnson Space Center Oral History Project. http://www.jsc.nasa.gov/history/oral_histories.htm.

Bourland, Charles T., and Gregory L. Vogt. *The Astronaut's Cookbook: Tales, Recipes, and More.* New York: Springer, 2010.

Congressional Record. 111 Cong. Rec. 16514. Senate hearings, July 12, 1965.

Duke, Charlie, and Dotty Duke. *Moonwalker.* Nashville: Oliver-Nelson, 1990.

Flentge, Robert. L., and Ronald L. Bustead. "Manufacturing Requirements of Food for Aerospace Feeding." Technical Report of the USAF School of Aerospace Medicine, Brooks Air Force Base, Texas. SAM-TR 70-23. May 1970.

"Food for Space Is Studied at Brooks AFB." Military News Service, San Antonio, Texas. June 10, 1966.

Heidelbaugh, Norman D., and Marvin A. Rosenbusch. "A Method to Manufacture Pelletized Formula Foods in Small Quantities." Technical Report of the USAF School of Aerospace Medicine, Brooks Air Force Base, Texas. SAM-TR 67-75. August 1967.

Ingelfinger, Franz J. "Gastric and Bowel Motility: Effect on Diet." Paper presented at the Conference on Nutrition in Space and Related Waste Problems, Tampa, Fla., April 27–30, 1964. Sponsored by NASA and the National Academy of Sciences.

Lepkovsky, Samuel. "The Appetite Factor." Paper presented at the Conference on Nutrition in Space and Related Waste Problems, Tampa, Fla., April 27–30, 1964. Sponsored by NASA and the National Academy of Sciences.

Murphy, Edwin, L. "Flatus." Paper presented at the Conference on Nutrition in Space and Related Waste Problems, Tampa, Fla., April 27–30, 1964. Sponsored by NASA and the National Academy of Sciences.

Slonim, A. R., and H. T. Mohlman. "Effects of Experimental Diets and Simulated Space Conditions on the Nature of Human Waste." Technical Report of the Aerospace Medical Research Laboratories, Wright-Patterson Air Force Base, Ohio. AMRL-TR 66-147. November 1966.

16 EATING YOUR PANTS

Kleiber, Max. "Animal Food for Astronauts." Paper presented at the Conference on Nutrition in Space and Related Waste Problems, Tampa, Fla., April 27–30, 1964. Sponsored by NASA and the National Academy of Sciences.

Worf, D. L. "Multiple Uses for Foods." Paper presented at the Conference on Nutrition in Space and Related Waste Problems, Tampa, Fla., April 27–30, 1964. Sponsored by NASA and the National Academy of Sciences.